EXPERIMENT
AND
STATISTICAL BASIS

THE SCIENTIST'S LIBRARY
Biology and Medicine

EDITED BY
PETER P. H. DE BRUYN, M.D.

EXPERIMENTAL DESIGN
AND ITS
STATISTICAL BASIS

D. J. FINNEY

THE UNIVERSITY OF CHICAGO PRESS
CHICAGO & LONDON

THE UNIVERSITY OF CHICAGO COMMITTEE
ON PUBLICATIONS IN BIOLOGY AND MEDICINE

EMMET B. BAY · LOWELL T. COGGESHALL
LESTER R. DRAGSTEDT · FRANKLIN C. McLEAN
THOMAS PARK · WILLIAM H. TALIAFERRO

The University of Chicago Press, Chicago 60637
The University of Chicago Press, Ltd., London

*Copyright 1955 by The University of Chicago. All rights reserved
Published 1955. Fifth Impression 1974. Printed in the United
States of America*

*International Standard Book Number: 0-226-24999-9 (clothbound)
Library of Congress Catalog Card Number: 55-10245*

I *will also tell you of an experiment that has been made in this kingdom of Kerman.*

The people of Kerman, then, are good, very humble, peaceful, and as helpful to one another as possible. For this reason, one day that the King of Kerman was surrounded by his wise men, he said to them: "Gentlemen, I am greatly astonished at not knowing the reason of the following fact: namely that, whereas in the kingdoms of Persia, so near to our land, the people are so wicked and treacherous that they constantly kill one another, with us, who yet are almost one with them, there hardly ever occur outbursts of wrath or disorder." The wise men answered him that the cause lay in the soil. Then the King sent some of his men into Persia, and particularly to the Kingdom of Isfaan above mentioned, whose inhabitants surpassed all the others in wickedness. Here, on the advice of his wise men, he had seven ships loaded with earth, and brought to his kingdom. When the earth arrived, he had it sprinkled, after the manner of pitch, on the floor of certain much frequented rooms; and had it covered with carpets, in order that its softness should not soil those present. There they then sat down to a banquet, and straightway, at the very first course, they began offending one another with words and deeds, and wounding one another mortally. Then the king declared that truly the cause of the fact lay in the soil.

The Travels of Marco Polo

A mighty maze! but not without a plan.

Alexander Pope, *Essay on Man*

Preface to the Series

During the past few decades the investigative approaches to biological problems have become markedly diversified. This diversification has been caused in part by the introduction of methods from other fields, such as mathematics, physics, and chemistry, and in part has been brought about by the formulation of new problems within biology. At the same time, the quantity of scientific production and publication has increased. Under these circumstances, the biologist has to focus his attention more and more exclusively on his own field of interest. This specialization, effective as it is in the pursuit of individual problems, requiring ability and knowledge didactically unrelated to biology, is detrimental to a broad understanding of the current aspects of biology as a whole, without which conceptual progress is difficult.

The purpose of "The Scientist's Library: Biology and Medicine" series is to provide authoritative information about the growth and status in various areas in such a fashion that the individual books may be read with profit not only by the specialist but also by those whose interests lie in other fields. The topics for the series have been selected as representative of active fields of science, especially those that have developed markedly in recent years as the result of new methods and new discoveries.

The textual approach is somewhat different from that ordinarily used by the specialist. The authors have been asked to emphasize introductory concepts and problems, and the present status of their subjects, and to clarify terminology and methods of approach instead of limiting themselves to detailed accounts of current factual knowledge. The authors have also been asked to assume a common level of scientific competence rather than to attempt popularization of the subject matter.

Consequently, the books should be of interest and value to workers in the various fields of biology and medicine. For the teacher and investigator, and for students entering specialized areas, they will provide familiarity with the aims, achievements, and present status of these fields.

PETER P. H. DE BRUYN

CHICAGO, ILLINOIS

Foreword

This book is an attempt to outline the theory and practice of that branch of statistical science generally known as *experimental design*, in a form that will be intelligible to students and research workers in most fields of biology. I have emphasized the basic logical principles and the manner in which their application aids the investigation of specific problems of research in medicine, genetics, pharmacology, agriculture, biochemistry, and other branches of pure and applied biology; I have deliberately neglected technical details of the theory, except to the extent that brief comments on these are essential to development of the theme. Even a reader who lacks both mathematical ability and acquaintance with standard methods of statistical analysis ought to be able to understand the relevance of these principles to his work, if he will devote some hours to their critical study. He may not comprehend the full reasons for all practices of experimental design, but he should gain a new outlook on his own experimentation that will prove of far greater value than any purely mathematical skill in the arithmetic of statistical analysis. To him this book is addressed, with no intention that it shall act as a textbook but in the hope of arousing his interest in a subject whose importance to good research practice is increasingly recognized. I make no claim that the subject is easy, but only that those who will rid themselves of the fear

Foreword

of mathematics can understand much without using advanced mathematical techniques.

I am grateful to Dr. H. Kalmus for permitting me to use unpublished details of his experiment discussed in chapter vi. I am also glad to express my thanks to Dr. M. R. Sampford and to my father, Mr. Robert G. S. Finney, for valuable comments on the text, and to Mrs. D. M. Russen for her continued patience in typing successive drafts.

D. J. FINNEY

OXFORD

Contents

I. STATISTICAL SCIENCE 1
II. COUNTS 9
III. MEASUREMENTS 29
IV. RANDOMIZED BLOCKS AND LATIN SQUARES 45
V. INCOMPLETE BLOCK DESIGNS 68
VI. FACTORIAL EXPERIMENTS 82
VII. SEQUENTIAL EXPERIMENTS 113
VIII. BIOLOGICAL ASSAY 123
IX. THE SELECTION OF A DESIGN 142
REFERENCES 162
INDEX 167

CHAPTER I

Statistical Science

1.1. WHY STATISTICS?

Shortly after detergents in powder form for domestic use first appeared on the British market, my wife remarked to a friend that she found a particular brand very good for clothes washing. "I would never use that," said her friend, in a horrified tone, "why, it's a chemical!" Despite increasing realization that many of the problems of biological science are intrinsically statistical, "why, it's statistical!" probably remains the unspoken reason for many biologists neglecting to employ techniques that could in reality aid their research. The notion that experiments and other research investigations can be conducted statistically or nonstatistically, at the will of the investigator, is firmly held by many: it is usually entirely false.

The biologist who wishes to record what he has observed must choose between descriptions, counts, measurements, and some combination of these three. A taxonomist may describe a new species of insect, with particular reference to differences from other species of the same genus; a geneticist may count the numbers of seedlings from the crossing of two selected parents falling into different categories; a clinical research worker may record how many of his patients are in various stages of recovery six weeks after a specific course of treatment was begun; a biochemist may record the weights of various organs of rats that have received different diets.

Moreover, one characteristic common to all biological material is that it varies: if a sufficiently discriminating measuring instrument is used, animals or plants treated alike will differ in respect of very many measurable properties. When observations are recorded as counts, not only may this variation sometimes lead to uncertainty in the classification of certain individuals, but individuals alike in origin and treatment may differ in their classification. Even though only a description of certain individuals or phenomena is wanted, this is superficial unless it takes account of the range of variation encountered in a group broadly classified as similar. When several groups of observations, arising from material subjected to different treatments or collected from different sources, are to be compared, any real differences will be to some extent masked by this variation. On the other hand, what appears to be a genuine difference attributable to the contrast of alternative treatments may perhaps be due, wholly or in part, to the chance conjunction of natural variations.

Such records are, by their very nature, statistical, and many of the inferences that a biologist would wish to draw from them depend upon statistical modes of thought. For example, if an experimenter weighs two sets of eight rats, whose history has been the same except for a difference in one component of diet, and if he asserts the greater mean weight of one set to be a consequence of its diet, unless he is being very naïve, he is making both a statistical inference that the difference is too great to be attributable to chance variations between individual rats and a logical inference that causes other than the contrast of diets can be excluded.

In brief, the biologist concerned with any quantitative assessments must use statistical methods, whether or not he gives them that name. His only choice is between good methods and bad, between methods with a sound theoretical basis that are appropriate to the problem and those that are untrustworthy or irrelevant; too often a wrong choice follows from failure to appreciate the statistical character of a problem or from attaching excessive importance to simplicity of method.[1]

Even those who know the need for careful statistical analysis of their results are not always aware of the extent to which the quality of information obtainable from an experiment can be modified by various details of its conduct. This book is intended to provide an introduction to the principles and potentialities of *experimental design*, in a form that can be understood by biologists with no special training in statistics or mathematics. With this limitation and in so short a volume, a comprehensive account of methods of design and the analysis of results is impossible; instead, the emphasis will be on illustrating the use of a wide variety of designs and discussing the broad principles to be followed in planning experiments.

1.2. EXPERIMENTAL DESIGN

By the "design" of an experiment is meant: (i) the set of treatments selected for comparison; (ii) the specification of the units (animals, field plots, samples of blood) to which the treatments are to be applied; (iii) the rules by which the treatments are to be allocated to experimental units; (iv) the specification of the measurements or other records to be made on each unit. The relevance of an experiment to the problems under investigation and the trustworthiness of conclusions drawn from the experiment depend very largely upon these matters. Moreover, all are to some extent the

[1] I have often observed that biologists tend to select textbooks of statistical methods almost entirely on the criterion of easiness to read. Desirable as this trait is, it would scarcely be regarded as a sufficient guide to authoritative information on any other science!

concern of statisticians. Although the set of treatments is largely the responsibility of the experimenter, statistical theory contributes ideas on the optimal choice (see especially chaps. vi, viii, ix). The experimenter, who may have little freedom of choice, often selects his experimental units unaided, but statistical analysis of past records can be valuable in indicating what specifications are likely to give the most precise results: questions of the age at which animals are most sensitive to differences of treatment, the dimensions of field plots that will enable yields of crops to be validly and precisely measured, or the dilution of a suspension of cells that will permit the most accurate counts to be made enter here, and the answers almost invariably depend upon detailed analysis of previous similar experiments. What might almost be termed the "classical" theory of experimental design is briefly described as the system of rules for allocating treatments to experimental units. In the past, this has been the aspect of design most studied by statisticians, and it forms the main theme of this book. The most important records to be made are those directly used in the evaluation of the treatments. Their general character is determined by the nature of the experiment, but statistical considerations enter into decisions on such matters as the number of plants on a plot that are to be examined for insect damage, or the size of a blood sample, the time that elapses between treatment and taking the sample, and the number of independent cell counts to be made on the sample. In addition, records of other characteristics of the experimental units (initial weight of an animal or physical and chemical properties of the soil of plots) that are necessarily unaffected by the treatments but may influence responses to treatments can be valuable as concomitant information. The specifications of units and of records are considered a little more fully in chapter ix.

Experimental Design

Though scientists are sometimes reluctant to regard the planning of a research program in "pure" science in economic terms, they can never entirely escape economic considerations. In applied science, the limiting factor to a program may be the total monetary expenditure. In pure science the monetary control may be less obvious, at least for work that is to form part of the normal activity of a laboratory or research team, but supplies of subjects or materials may be equally effective limitations; even when this does not obtain, the program will be limited by the total time that can be spared for it among the competing claims of alternative lines of research. Whatever the limiting factor, it is obviously desirable to consider, before experiments are begun, how resources can be used most advantageously. On matters so fundamental to the nature and conduct of an experiment as those listed at the beginning of this section the statistician is no final arbiter. He should, however, give important help in eliminating sources of bias that might lead to false inferences and in insuring that resources are so utilized as to produce the most precise estimates of numerical quantities and the most sensitive tests of hypotheses (see § 2.6 for a simple example). A further gain from good experimental design is that often the conclusions to be drawn are so patent as to make laborious statistical analysis scarcely necessary.

If the best results are to ensue, close collaboration between experimental scientists and statisticians is essential, for neither can design experiments well without understanding the point of view of the other. "Statistical science is one of the precision instruments available to the experimenter, who, if he is to make proper use of the knowledge at his disposal, must either learn to handle it himself or find someone else to do so for him. Experimenters who will put themselves to great trouble in acquiring skill with some difficult biological or chemical technique often deny themselves the benefits of statistical techniques because they consider these beyond their understanding. The fault may lie in part with statisticians, in that they fail to make their methods sufficiently clear to the non-mathematician, but the loss is entirely the experimenters'" (Finney, 1952b). This book is written in the belief that the principles of experimental design and their more important practical applications can be appreciated by any scientist, however restricted his formal training in mathematics.

1.3. BOOKS ON STATISTICAL METHOD

There are today numerous good books that instruct the reader in statistical methods for use in the biological sciences. The choice between them rests largely on personal taste and field of interest, and no list is given here. Section A of the References (p. 162) contains the titles of several exceedingly elementary introductions to the methods of statistical science. In the main, these are less concerned with the practice of the methods than with describing and illustrating the basic principles; they provide information on statistical methods complementary to that given here on experimental design. Quenouille, Snedecor, and, in a more specialized field, both Hill and Bernstein & Weatherall are also useful texts of elementary methodology.

1.4. BOOKS ON EXPERIMENTAL DESIGN

Section B of the References (p. 162) lists the more important books on the theory and practice of experimental design. Fisher's book is pre-eminent for explanation of the philosophy of design without details of theory. Cochran and Cox provide an excellent manual of instruction on how to deal with standard designs, especially those outlined in chapters v and vi, and Yates gives similar information in more condensed form; Quenouille has more to say about the choice of designs and the interpretation of results, less about details of analysis. Davies gives very detailed instructions and examples but is nonbiological. Kempthorne's book is the most comprehensive treatise yet available on the theory of designs. Cochran and Cox's extensive catalogue of designs can usefully be supplemented by Fisher and Yates's tables that are in any case almost essential to the biologist who uses statistical techniques, because they include the standard χ^2, t, variance ratio, and other important tables regularly wanted in statistical analysis. Kitagawa and Mitome have given an even fuller catalogue of designs, displayed in Roman characters with a long accompanying text in Japanese.

1.5. THIS BOOK

This book is less ambitious than any mentioned in § 1.4. It is not a manual of instruction on the design and analysis of experiments but a general survey of how statistical theory can usefully guide experimental design. Written entirely for biologists, it assumes no previous knowledge of statistical practice. However, since experimental design can scarcely be understood in ignorance of the manner in which the results of experiments are analyzed, a few basic ideas on statistical analysis are explained in the early chapters, the most important being related to contingency tables and the analysis of variance; though not essential, acquaintance with one of the books in Section A of the References will help the reader. No knowledge of mathematics is needed beyond the ability to comprehend a few algebraic symbols: much of chapters v and vi relates to combinatorial mathematics, but all can be understood, without special mathematical theory, by anyone who will take a little trouble.

The book is planned for consecutive reading, rather than as a work of reference, but a reader who finds difficulty with some sections of chapters v and vi will perhaps do well to continue with subsequent chapters before struggling greatly with their difficulties. Section C of the References (p. 163) records books and papers mentioned in the text but is in no way a comprehensive bibliography of the theory and practice of experimental design. References have been given only when the text uses other published work as the source of illustrations or where particular papers seem likely to help readers for whom the book is planned.

CHAPTER II

Counts

2.1. Records of Frequencies

In many experiments, the observations are recorded as counts of "events" or occurrences in different categories. The simplest case is that in which only two categories are recognized: black and white, dead and alive, male and female, or normal and diseased. More elaborate classifications are encountered, however, such as the frequencies of insects dead, moribund, and recovered after exposure to an insecticide or the frequencies of cases of cancer at different sites among men who have also been classified according to their smoking habits.

The interest of geneticists in the qualitative classification of individuals has the result that records of genetical experiments are especially often of this type. Since some of the most readily appreciated applications of statistical techniques relate to the examination of genetical theories of the frequencies with which alternative genotypes and phenotypes occur, one or two examples can well begin this chapter. Unfortunately, this theme cannot be developed very far; the statistical methods required become more difficult and highly specialized so rapidly that an account of the design of experiments in genetics would need a separate book.

2.2. Deviations from a Theoretical Proportion

In a paper on the genetics of the alpine poppy, Fabergé (1943) reports (among many other results) segregations ob-

Counts

tained by backcrossing plants with green bases to purple-based parents. One family of 28 seedlings was classified as 9 purple, 19 green. If this were the only evidence available, would it be consistent with the hypothesis that the green base is determined by a simple recessive gene, v, the purple parent being heterozygous Vv? Simple Mendelian theory states that individual seedlings from this backcross are as likely to be purple as green, so that progenies of 28 should average 14 of each.[1] At first sight, the family under discussion appears to have a marked excess of green. However, if many families of 28 were grown, some would have more and others less than 14 purples, and the question propounded is therefore equivalent to inquiring whether so large a deviation as that recorded can reasonably be attributed to chance variations from family to family.

If the hypothesis is correct, each seedling produced is as likely to be purple as green, in just the same sense that (ideally) a well-balanced coin, spun fairly, would be as likely to show heads as tails. Hence the relative rarity of a family as extreme as this might be investigated by spinning a set of 28 coins many times and seeing how often the deviation of the numbers of heads and tails from equality is as great as 9 to 19. If such occurrences were very rare, it would be reasonable to infer that the hypothesis was false; if they were fairly common, it would be clear that even this apparently large deviation could easily arise by chance and was therefore little evidence against the hypothesis. Thus a trial with 28 coins simulates the behavior of the genetical experiment, with the advantage that it can be repeated many times in order to build up empirical knowledge of the *frequency distribution* of the number of heads (or the number of purples) in families

---

1. Using the words in a special sense, the statistician calls 14 the *expected number* or the *expectation* in each class.

of 28. This approach is laborious, however, since thousands of trials would be necessary in order to determine the distribution at all satisfactorily, and fortunately a simple mathematical approach can be used instead. Since each coin has two possible positions, "head" and "tail," and every possibility for one coin can occur in combination with every possibility for each other coin, the total number of possible results is 2^{28} (about 2.7×10^8), all being equally likely to occur. Of these, 1 has 28 heads, 28 have 27 heads and 1 tail, 378 have 26 heads and 2 tails, and so on.[2]

By direct calculation in this way, or by reference to the published tables mentioned below, the proportion of results in which the number of heads is 9 or less is found to be 0.0436. Hence, in a long series of trials, the relative frequency of "9 heads or less" among all the 2^{28} equally likely possibilities is 0.0436. This is known as the *probability* of results in that category, and, because the coin experiment is a model of the genetic experiment, it is also the probability of finding 9 or less purples in the family of 28 if the hypothesis of recessivity of the green condition be correct. In assessing the strength of the evidence against the hypothesis, however, we must remember that a deviation from average in the opposite direction, 19 or more purples and therefore 9 or less greens, would have been equally potent, and symmetry shows the probability of this also to be 0.0436. The total probability of a deviation from perfect agreement with the hypothesis as great as or greater than that observed is thus 0.087: even though the hypothesis of a simple recessive gene for green were correct, about 1 family in 11 of the same parentage and size would deviate from equality of the two classes as

2. The number of results having exactly r tails is the numerical coefficient of x^r obtained when $(1 + x)^{28}$ is multiplied out completely. Proof is not difficult; most readers will satisfy themselves that it is true by verifying the corresponding statement for a small number of coins (3, 4, 5) from counts of all possible cases.

markedly as does this one. Hence this family can scarcely be considered to provide much evidence against the hypothesis.

This type of test can be applied to other segregations. For example, an F_2 progeny raised in the same investigation showed 10 purples and 8 greens; on the hypothesis that the color is determined by a simple recessive, both parents are Vv, and individual seedlings have a chance of $\frac{3}{4}$ of being purple, $\frac{1}{4}$ of being green. Again there appears to be a deficiency of purples. A model could be set up by spinning 18 pairs of coins, one pair for each seedling; a pair of coins that shows at least one head corresponds to purple, and a pair that shows two tails to green. Again actual trial with coins could be made the basis of an investigation into the rarity of a deficiency of purples as extreme as that observed, and again a simple mathematical approach is easier and quicker.[3] By direct calculation or from tables, the probability of getting 10 purples or less (as compared with the expectation of $13\frac{1}{2}$ predicted by the hypothesis) is 0.057. Although for unequal probabilities in the two classes there are no precisely corresponding deviations in the opposite direction, allowance must still be made for the possibility of a chance excess of purples by doubling this value; thus 0.114 is taken as the total probability of deviations as great as or greater than that observed. Since so large a deviation would arise by chance about once in 9 times, this family also constitutes no great evidence against the hypothesis.

On the other hand, if the F_2 family had contained 8 purples and 10 greens, the similarly calculated probability would have been 0.011, a much stronger indication of a flaw in the underlying hypothesis. Although some experiments can lead

3. There are 4^{18} (2^{36}, or 6.9×10^{10}) possible arrangements of heads and tails among the 36 coins, and the numerical multiplier of x^r in $(3 + x)^{18}$ is the number of these in which r of the 18 pairs consist of two tails.

to the total rejection of a hypothesis on the basis of a critical observation (except for the possibility of a mutation, the occurrence of a single purple among the progeny of a cross between two greens would disprove the hypothesis that green was a simple recessive character), often a decision must rest upon assessment of probabilities, and the experimenter can regard a hypothesis as disproved only because its truth would assign a very small probability to the observations. He is free to choose what value he likes as the "very small probability" for a particular experiment, provided that he chooses before he knows the results, and he will rightly take a larger value if he is particularly anxious not to miss any indications of departure from hypothesis than if he is interested only in a departure so large and unmistakable that the importance of acting upon it is undeniable. In many fields of quantitative biology, it has become customary to speak of a probability of 0.05 or less as providing *statistically significant* evidence against the hypothesis on which its calculation was based and as justifying rejection of this hypothesis. Nevertheless, the convention of using the word *significant* as meaning a probability of 0.05 or less, and similarly *highly significant* for 0.01 or less, is in no way an absolute standard: whenever an alternative (e.g., 0.1 or 0.001) seems more appropriate to particular circumstances, it should be used unhesitatingly—of course with the change from convention clearly stated.

In this manner, observed counts in any two categories can be compared with proportions specified by genetic hypotheses or other theoretical considerations. Always a model for repeated trials can be set up,[4] and always arithmetical processes can be used in direct computation of probabilities according

4. For example, if hypothesis stated a proportion of $\frac{1}{3}$ in one category, results of throwing a standard cubical die could be used; two of the six faces would be taken to correspond to this category and four to the other.

to the *binomial distribution*, of which examples have been given. Tables have been prepared from which the probabilities can be read directly for small numbers (National Bureau of Standards, 1950), and for larger numbers the χ^2 approximation (§ 2.3) is usually sufficiently accurate.

2.3. The χ^2 Distribution

If theory states that a fraction P of observations ought, on an average, to fall into one of two classes, and of a set of n independent trials the proportion in this class is p, then the quantity χ^2 ("chi-squared"), defined by

$$\chi^2 = \frac{n(p-P)^2}{P(1-P)},$$

can be used to approximate to the test of significance of the deviation of p from the theoretical value. Provided that nP and $n(1-P)$ are fairly large, the probability that χ^2 exceeds any specified value is practically independent of n and P; if chance alone is responsible for the deviation from P, the probability that χ^2 exceeds 3.84 is 0.05, and the probability that it exceeds 6.63 is 0.01.

For example, the F_2 progeny containing 10 purples and 8 greens, discussed in § 2.2, has for the proportion of purples

$$P = 0.75, \quad p = 0.556 \text{ (i.e., } \tfrac{10}{18}\text{)}.$$

Using the adjustment mentioned in footnote 5,

$$\chi^2 = \frac{18 \times (0.750 - 0.556 - 0.028)^2}{0.75 \times 0.25}$$
$$= 2.65.$$

Since this is less than 3.84, it does not exceed the 0.05 significance level; reference to more detailed tables of χ^2 (Fisher and Yates, 1953) assigns it a probability 0.104, which approximates to the exactly calculated 0.114. The use of χ^2 is in fact rather unsafe when nP or $n(1-P)$ is small (say less than 5) and may then give only a poor approximation,[5] but, when applicable, it saves much arithmetic.

2.4. Counts in More than Two Classes

If the objects counted fall into more than two classes (as often occurs with genetical observations), an extension of the χ^2 method enables the deviation from hypothesis to be tested. Any good textbook of statistical science gives details.

2.5. Disproof, Proof, and Estimation

To conclude that certain observations do not disprove a hypothesis does not amount to proof of the hypothesis, a statement that is readily apparent but frequently forgotten. The observation of 10 purples and 8 greens is obviously consistent with a 1:1 segregation as well as with a 3:1. If no specific genetic hypothesis were in mind, the experimenter might wish to estimate what proportion of greens this mating would, on an average, produce. Clearly, his estimate that the average in a long series would coincide with the value from the experiment, $\frac{8}{18}$ or 0.44, will not necessarily be exactly correct; indeed, the test of significance already described permits any theoretical ratio to be tested and so provides a method of determining what values are rejected and what are not. By testing a series of values of P exactly as in § 2.2, he will find that the only ones not rejected by the test of significance are those between 0.22 and 0.64, and these extremes may therefore be regarded as limits of error: they are in no sense absolute limits, but, if in similar problems limits are habitually so assessed, the statement that the true value lies between the limits will usually be correct. If

---

[5]. In general, the approximation is improved by subtracting $\frac{1}{2n}$ from the difference between p and P before squaring (*Yates's Continuity Correction*). Many different but equivalent formulae for χ^2 are in use, and that given here is not always the most convenient for computing.

the 18 seedlings had been a random selection of unrelated individuals from a random-mating population for which the recessivity of green was known, the ratio $\tfrac{8}{18}$ would estimate the relative frequency of recessive individuals in the population. As is well known from the theory of random mating in population genetics (Hardy, 1908; Stern, 1950), if the population is in equilibrium, this quantity is the square of the gene frequency; by taking square roots throughout, the frequency of the v gene is then estimated at 0.67 and asserted to lie almost certainly between 0.47 and 0.80.

2.6. THE PLANNING OF GENETICAL EXPERIMENTS

Suppose that, in a situation such as that discussed in § 2.2, there were reason to suspect incomplete penetrance of the v gene, a proportion θ of all vv homozygotes being purple and thus phenotypically indistinguishable from the other two genotypes. Then the average relative frequency of purples from a backcross should be

$$P = \frac{1+\theta}{2}$$

and, from F_2

$$P = \frac{3+\theta}{4}$$

instead of $\tfrac{1}{2}$ and $\tfrac{3}{4}$, respectively. Hence, from a sample of plants classified to give an estimate p of P,

$u = 2p - 1$ for a backcross   and $u = 4p - 3$ for an F_2

estimates the unknown quantity θ. Now inspection of the formula for χ^2 in § 2.3 indicates that $P(1-P)/n$ is a measure of the extent to which p is likely to vary about P in a progeny of size n; this is apparent because the probability associated with any particular value of $(p-P)^2$ is dependent only on $(p-P)^2 \div P(1-P)/n$, so that the divisor scales down any squared deviation $(p-P)^2$ in such a way as to eliminate the influence of P and n on its probability. For instance, the probability that p differs from P by more than $1.96\sqrt{[P(1-P)/n]}$ is 0.05 (since $1.96 = \sqrt{3.84}$). In fact, $P(1-P)/n$ is the *variance* of p, and its square root the *standard error* of p (§ 3.3). Moreover, the variation to which u is subject is obviously twice that for P for the backcross, four times for the F_2. When written in terms of θ, the formula becomes, for the backcross,

$$\text{Standard error of } u = \sqrt{\frac{1-\theta^2}{n}},$$

and, for the F_2,

$$\text{Standard error of } u = \sqrt{\frac{(3+\theta)(1-\theta)}{n}}.$$

For every possible value of θ (of course less than 1) the second standard error is greater than the first, so that the backcross is always more informative. In particular, if penetrance is almost complete and θ therefore very small, estimates of θ from F_2's will be subject to almost $\sqrt{3}$ times the variation that estimates from backcross progenies of the same size would show: in other words, to obtain a standard error for an F_2 as small as that from a backcross progeny of n, $3n$ individuals would be needed. For detecting incomplete penetrance, backcrosses are much more useful than F_2's, being in fact three times as sensitive to disturbances of the segregation. More generally, the *efficiency* of backcrosses relative to F_2's is

$$\text{Efficiency} = \frac{(3+\theta)(1-\theta)}{1-\theta^2} = \frac{3+\theta}{1+\theta},$$

so that even if θ were almost unity (the recessive vv almost completely failing to manifest itself), each member of a backcross progeny would be twice as useful as a member of an F_2 progeny in the estimation of θ.

This discussion relates only to deviations from Mendelian

ratios attributable to incomplete penetrance, and deviations due to other causes would lead to different assessments of the relative efficiency of the two types of progeny. An analysis of the same general character can be applied to these and to more complex genetical problems in order to determine the most efficient experimental procedure for a particular purpose.

2.7. THE COMPARISON OF PROPORTIONS

When two (or more) proportions are to be compared, a slightly more complicated analysis is required. Table 2.1

TABLE 2.1

MORTALITY FROM MYOCARDIAL INFARCTION

Treatment	Survived	Died	Total	Per Cent Mortality
Control	74	51	125	41
Anticoagulant	56	19	75	25
Total	130	70	200	35

shows results reported in a study of anticoagulant therapy for myocardial infarction (Loudon *et al.*, 1953); these will be discussed somewhat uncritically in the first place, purely as an illustration of statistical technique, after which the relationship of the analysis to the interpretation of the results will be considered.

Here the interest no longer lies in comparing a proportion with a value predicted by some hypothesis, but in examining the strength of the evidence that anticoagulant therapy alters the mortality rate from its value among control subjects not receiving the therapy. The control mortality rate is not specified by any theory but can be estimated from the first line of Table 2.1, and the first question to be answered is whether the mortality in the second line is consistent with

Comparison of Proportions

a belief that the same rate operates. A *null hypothesis* may be stated: "The two groups of patients are subject to identical death rates, and differences in the proportions actually dying are due solely to chance variations"; the extent to which Table 2.1 provides evidence against this hypothesis must then be assessed.

If the null hypothesis be true, the third line of the table provides an estimate of the over-all death rate from myocardial infarction. The total number of deaths, 70, itself tells nothing of any difference in rates between the control and treated groups, and whatever information the table gives must lie in the division of these 70 between the two groups. A model can be set up by taking 200 pieces of white card, all alike except that 75 bear a distinguishing red mark, so corresponding to the patients receiving anticoagulant, whereas the remainder correspond to the controls. After thorough mixing, a sample of 70 is drawn (to correspond to deaths) and its members are classified as "white" or "red." The sample is then mixed with the other cards and a new sample drawn. Repetitions of this process lead to empirical construction of the relative frequencies with which the 71 possible classifications occur (70 white; 69 white and 1 red; 68 and 2; . . . ; 1 and 69; 70 red), and, if a large number of trials is made, these will approximate to the probabilities of the classifications under the condition of the null hypothesis. Thus the probability of obtaining results in which the observed difference in the proportions dying in the two groups was at least as great as in Table 2.1 could be found. Once again the empirical process can be replaced by an arithmetical one, using the fact that the number of deaths in each group must follow a binomial distribution; but even this is somewhat laborious for numbers as large as those in Table 2.1.

Counts

Fortunately the χ^2 distribution (§ 2.3) again provides a good approximation. The first step is to calculate the deviations of the observed frequencies in the two groups from perfect agreement with the proportion shown by the totals. If the deaths in the control group had agreed perfectly with the over-all proportion, 70 out of 200, the second entry in Table 2.1 would have been 43.75 ($= 125 \times 70/200$), and the other entries would have been similarly modified. Table 2.2 shows these expectations and the deviations of the entries in Table 2.1 from them.[6] The deviations must be equal in size and two of each sign, and their magnitude is a measure of the

TABLE 2.2

EXPECTATIONS CORRESPONDING TO TABLE 2.1
(Deviations in Parentheses)

Treatment	Survived	Died	Total
Control..............	81.25(+7.25)	43.75(−7.25)	125(0)
Anticoagulant........	48.75(−7.25)	26.25(+7.25)	75(0)
Total............	130 (0)	70 (0)	200(0)

extent to which the frequencies disagree with the null hypothesis: χ^2 is found by reducing the deviations by $\frac{1}{2}$, squaring each and dividing by the corresponding expectation, and adding the four quotients:

$$\chi^2 = \frac{(6.75)^2}{81.25} + \frac{(-6.75)^2}{43.75} + \frac{(-6.75)^2}{48.75} + \frac{(6.75)^2}{26.25}$$
$$= 0.56 + 1.04 + 0.93 + 1.74$$
$$= 4.27 .$$

The card-drawing model would require entirely different calculations for every different set of totals in Table 2.1; the

---

6. 43.75 deaths would be a strange phenomenon to observe! So would a family of 2.37 children: nevertheless, 2.37 children might be the average size of family in some community, and for exactly the same reason the technical sense of "expectation" permits fractions of individuals to occur.

remarkable fact is that the probabilities associated with different values of χ^2 are almost independent of these totals, provided only that the frequencies under examination are not unduly small.[7] The probabilities associated with χ^2 are the same as in § 2.3. Hence, since the calculated value exceeds 3.84, the null hypothesis is rejected on the basis of a test of significance at the 0.05 level, and the difference in death rates for the two groups is statistically significant.

2.8. INTERPRETATION OF A SIGNIFICANCE TEST

Significant of what? That question must always be asked. The investigator here would like to conclude "significant of an improvement arising from the use of anticoagulant therapy," but such an answer can be made with confidence only if other explanations can be ruled out, which in turn requires that the two groups shall be comparable in every other way. In this investigation, 75 of the control cases occurred between 1945 and the introduction of anticoagulant therapy into the hospital in 1950; from then until May, 1952, use of anticoagulants was determined by the views of the patient's physician. Any improvement in other conditions, having no causal connection with the new treatment, could therefore be reflected in a lower mortality rate in the later period, but, since the deaths in the two periods among the controls amounted to 39 and 44 per cent, there seems to be little sign of this. Again, heterogeneity of the origins of the records might have important effects. Any tendency for patients to be assigned to the anticoagulant group when their condition was less serious and offered better chances of recovery would obviously bias the results in favor of this treatment. Sex or age differences in mortality rates could produce

[7]. A rough rule is to use the χ^2 test only if every expectation exceeds 5. Many people calculate χ^2 from the squares of the actual deviations, but reducing these by $\frac{1}{2}$ before squaring improves the approximation.

apparently better results for the new treatment if the representation of the sexes or ages in the two groups differed appreciably. The physicians who favor the use of anticoagulant therapy may have been "better" than the others in ways entirely unconnected with this treatment, but the consequences would appear in Table 2.1 as indistinguishable from direct effects of treatment.

When faced with Table 2.1, the statistician can do little but demonstrate the existence of a significant difference and suggest possible explanations; he is not himself competent to controvert explanations on which no objective information exists. One of the difficulties of research in clinical medicine is that restrictions on the manner in which experiments can be performed may prevent the logical exclusion of explanations of the results other than the one that is wanted. After careful examination of all available evidence, Loudon and his colleagues concluded that none of the factors just mentioned was likely to have played any important part and that therefore the significant difference in mortality rates must be an effect of anticoagulant therapy. In this instance as in many others, those who have been closely associated with the investigation are doubtless best qualified to discuss the alternatives, but inevitably least able to eliminate subjective judgment. If consulted at an early stage, however, the statistician can sometimes make a contribution to a clinical investigation that will simplify the eventual drawing of right inferences. He does this through careful attention to the design of the experiment (§ 1.2).

2.9. Experimental Design

Clinical experimentation involves ethical, human, and practical problems that are absent from much scientific research, and the example in § 2.7 has been deliberately

Experimental Design

chosen because the gravity of the issue accentuates these. In introducing ideas on experimental design, it is useful to disregard such difficulties for a moment and to consider how the experiment should be planned if it were concerned with plants or animals instead of with humans. A statistician would then advocate the following procedure.

Observations on control and treated subjects should be made contemporaneously, lest conclusions be biased by changes in conditions irrelevant to the general operation of the new treatment. Moreover, characteristics of the subject (such as age, sex, previous history) or of the severity of the disease must not affect the allocation to treatment. An investigator who is free to decide which of the two treatments a subject shall receive will almost inevitably allow his choice to be influenced, consciously or subconsciously, by his knowledge of the subject. If his judgment is sound, that may be excellent for the cure of the disease, but the experiment will be misleading if the control and treated groups are inherently different. An objective rule for allocation is therefore essential. One possibility is to allocate subjects alternately to the two groups: even this can produce a bias if the alternation is known to the man responsible for deciding whether or not a subject shall be included in the experiment, or is known to anyone who has opportunities of manipulating the order in which the subjects are presented! The only safeguard is *randomization*. Whether or not a subject receives the treatment must be decided by the fall of a well-balanced coin, the drawing of lots, or some similar random process. Although the random order can be prepared at the start of the experiment, its verdict in respect of any subject should not be disclosed in advance to anyone concerned with deciding which subjects shall be admitted to the experiment. Restrictions on the complete randomization of spinning a coin independently for each subject are permissible. For example, lots might be drawn in such a way as to restrict the total numbers of treated and untreated to specified values. This has the merit that the χ^2 or other test will be more sensitive to small differences in the true mortality rates if the numbers in the two groups are constrained to be about equal: if the experimenter is prepared to use 200 subjects, he is more likely to detect any real difference by putting 100 in each group than by putting 125 in one and 75 in the other. (If the treatment demands much greater expenditure of labor and materials per subject than does the control, resources would be used most efficiently by keeping more of the subjects as controls.) Although randomization tends to balance the two groups in respect of age of subject or other characteristics, further improvements are possible, for example, by restricting the numbers of treated and control subjects to equality in each of several age groups and not merely in the total. The experienced statistician will have in mind many possibilities, and the best for a particular investigation requires consideration of all the circumstances. An experimenter who is not himself experienced in design should therefore consult a statistician well before he begins the experiment; at this stage, records of previous experiments should be examined for any information they give on the relative merits of different designs and on special precautions that are necessary.

2.10. Design of Clinical Experiments

The logic of § 2.9 is as relevant to human subjects as to plants or animals, but the application of the arguments is often more difficult. In the study of relatively mild human ailments, such as headaches or colds, a randomized experiment along the lines described above may not present

Design of Clinical Experiments

great problems, especially since subjects may be induced to volunteer for tests of a new treatment. (Volunteers, of course, must still be randomly divided between control and treatment.) Precautions are needed if the assessment of a cure depends on any subjective judgment, or if faith in the efficacy of treatment may itself produce a cure. Schemes can be devised to prevent either physicians or patients from knowing who has had a new drug and who has had a superficially similar dummy treatment. Even nurses and others who have direct contact with the subjects may also need to be kept ignorant of the treatments, lest their comments should affect the patients' morale and lead the physician to make a biased judgment of the extent to which some minor ailment has been cured! An excellent illustration of this point, and of the need to have adequate controls, has been given by Jellinek (1946) in his comparison of three variants (A, B, C) of a standard remedy for frequent headaches. Fortunately he included in his experiment a *placebo* (D) as a control treatment. Four sets of 50 subjects (one of which was later reduced to 49) received the remedies in successive fortnightly periods, a different sequence of A, B, C, D being adopted for each group (according to a Latin square scheme: see § 4.10). For the 199 subjects, the mean success rates in the cure of headaches were

A	B	C	D
84%	80%	80%	52%

from which one is tempted to conclude that A, B, C do not differ in effectiveness to any appreciable extent. However, the successes with the *placebo* were restricted to 120 subjects, the other 79 reporting no cures. Table 2.3 summarizes these two groups separately. Clearly, all four materials showed about equal success rates in the first group, and, in view of the known pharmacological inactivity of D, it is hard to escape the conclusion that these subjects were suffering from psychogenic headaches that responded to "suggestion"; in the second group, the superiority of A to B and C is marked and is supported by more detailed statistical analysis. Jellinek rightly comments: "Banal as it may sound, discrimination among remedies for pain can be made only by subjects who have a pain on which the analgesic action can be tested."

TABLE 2.3

SUCCESS RATES IN THE CURE OF HEADACHES
(In Per Cent)

Patients Showing:	A	B	C	D
Cures with *placebo*....	82	87	82	86
No cures with *placebo*..	88	67	77	0

At the other extreme of difficulty is an investigation such as that on myocardial infarction. Up to 1950 the question of giving an anticoagulant did not arise, and from then on no physician who himself believed in the advantages of anticoagulant therapy would forego its use for certain of his cases in order to comply with an experimental program. This attitude is perfectly proper, for, as soon as a physician is convinced that a new treatment improves the chances of survival or cure for a patient, he must place his duty to do the best for his patients above the needs of experimental science. Nevertheless, medical research is not purely academic. The interests of both the general public and research workers lie in insuring that the superiority of good new treatments is demonstrated, that new treatments which are in reality bad or useless are detected and discarded before they become part of the tradition of medical practice, and that conclusions are based on trustworthy evidence efficiently obtained. In the history of a promising new treatment, there is likely to be a stage at which responsible opinion believes it to be comparable in effectiveness with the existing standard treatment but at which no one would confidently assert that it represented any real improvement. From then until accumulated evidence is strong enough to make either further use of the new treatment or further use of the old unethical, treatment of each new case is necessarily experimental, whether in fact the old or the new is used. As A. B. Hill has often emphasized, not only is this period an opportunity for planned experimentation, but failure to do a properly designed trial amounts to an unethical rejection of information that could be provided by subjects who are "necessarily experimental." Sequential experimentation (§ 7.4) provides a method of keeping the progress of an experiment continuously under review, and may prove to be an excellent method in some clinical problems.

Whatever the nature of a clinical trial, the statistician will rightly regard the principles of § 2.9 as ideals; an excellent detailed statement along the same lines by Hill (1951) should be read by all concerned with clinical research. The statistician must recognize, however, that overriding medical, ethical, or administrative considerations may compel some compromise with the ideals. Less stringent conditions of contemporaneity, homogeneity of subjects, and randomness can be accepted only with reluctance but will often be preferable to abandonment of the research. In the interpretation of results, careful attention must then be given to the extent to which the validity of conclusions could be affected by imperfections of design; usually the statistician can do no more than point out the dangers, leaving to the experimenter responsibility for assurances that they are unimportant.

This brief discussion necessarily oversimplifies complex questions. The chief difficulties in the planning of clinical trials are usually organizational rather than statistical. As in other branches of research, the statistician should be asked to collaborate from the start: although he may not encounter major theoretical problems, he may have difficulty in devising a design that is reasonably efficient yet does not conflict with rigidly imposed ethical and administrative constraints. If the gravity of decisions to be taken is greater than in other research, so much the greater is the need to plan the investigation for the avoidance of bias and for the elimination of subjective judgments about alternative explanations of the results.

CHAPTER III

Measurements

3.1. MEASUREMENTS

The distinguishing feature of observations in the form of counts is that they are necessarily whole numbers, which property of discreteness leads to the distinctive methods used in their statistical analysis. Measurements, on the other hand, whether they be of length, weight, time, or some derived quantity such as density or velocity, are not so restricted: however little two objects may differ in weight, one can always conceive of a third object having an intermediate weight. In practice, the limitations of measuring instruments interfere with the true continuity of scales. For example, records of weights determined by a balance that will weigh only to the nearest milligram are in reality counts of the numbers of objects that fall within ranges $\frac{1}{2}$–$1\frac{1}{2}$, $1\frac{1}{2}$–$2\frac{1}{2}$, $2\frac{1}{2}$–$3\frac{1}{2}$ mg., and so on. Except when a very coarsely scaled measuring instrument is used, this consideration can be ignored, and the methods of statistical analysis generally used for measurements are based upon an assumption of a continuous scale. For practical purposes, counts and measurements correspond to what the theoretical statistician recognizes as *discrete* and *continuous variates*.

Measurements usually convey more information about the objects measured than would mere classification and counting. Indeed, unless objects counted as members of a particular category are absolutely alike in respect of the character studied, a measurement of the degree to which they possess the character must be more informative than a simple statement of how many fall on one side or the other of a certain dividing line. The plants described in § 2.2 as having green or purple bases undoubtedly differed among themselves in the degree and extent of this coloring; to measure this, however, would increase both the labor of observing and the difficulty of interpretation, and the investigator rightly concentrated on the simple classification. To classify plants as "tall" or "short," instead of measuring individual heights, would obviously sacrifice much potentially useful information on height that could be fairly readily obtained; although this course might be justified in a genetic investigation where a sharp distinction between tall and short depended on segregations at a single locus and minor variations could be attributed to modifiers and environmental variation, it would scarcely be advocated, say, in a study of the effects of different levels of nutrition on height.

Records of measurements can easily be reduced to counts, a step that is sometimes useful in the interests of a rapid statistical analysis or for a provisional examination of results (§ 3.3).

3.2. Design for a Simple Experiment

An experimenter who is prepared to expend a specified amount of materials, time, and effort for a particular purpose will wish to proceed in such a manner as to obtain the best possible results from this expenditure. Alternatively, he may specify a degree of reliability (in a sense explained later) and wish to achieve this with a minimal expenditure of his resources. The two problems are essentially the same, since a design that is optimal for a specified expenditure must be such that no alternative could give equally good results for less expenditure, and the first is perhaps the easier to describe. In reality, the specifications are rarely absolutely rigid, but the simple statement is a convenient starting point to a discussion of experimental design.

Bacharach (1940) examined a claim that deprivation of vitamin E inhibits the storage of vitamin A in rats' livers. The design and analysis of one of his experiments can illustrate many important points.

Suppose that an experimenter is prepared to expend 20 rats and an appropriate amount of time and labor in one experiment on this question. He can assign some or all of the rats to a diet deficient in vitamin E, and, after a suitable interval, determine the vitamin A in the livers. Of one thing he can be certain: the amount of vitamin A in the liver will not be the same in every rat. Hence, if he is to have any indication of whether a measurement of vitamin A is small because of an individual peculiarity of a rat or because of the effect of vitamin E, he must put more than one rat on the diet. To put the whole set of 20 rats on the diet would give the most information on the level of vitamin A for this treatment and might seem the best policy, since the results could be compared with known values for normal rats. However, this would raise difficulties over the discovery of the "normal" records, for, even if measurements made previously in the same or another laboratory could be found, there would rarely be any assurance that they were in every way comparable except for the one dietary deficiency; almost inevitably, the experimenter would be unable to judge how far differences in vitamin A were due to differences in environment. The only safeguard is to make simultaneous trial of the deficient diet and the normal on comparable animals. A conflict then arises between the desire to have as many rats as possible on the deficient diet and the desire to have as many as possible on the control or standard diet with which the results are to be compared;

the compromise leading to the most precise comparison is the natural one of assigning equal numbers to each.

The simplest procedure is to select 10 rats entirely at random from the 20 and to assign these to the deficient diet. Strict randomness of selection, by drawing lots or by use of tables of random numbers (§ 4.5), is essential in order to remove the danger of initial inherent differences between the two groups. Any conscious effort to balance the two groups introduces a grave danger of subjective biases, unless an element of randomness is retained (as in § 3.6), and even attempts at haphazard selection can go seriously wrong. Those who have never put the matter to the test are often unaware of the difficulties in making a fair division into groups by subjective judgment or haphazardly, but many examples can be quoted of the way in which otherwise good experiments have been spoiled by failure to randomize. For example, the first 10 animals picked from a cage of 20 may have been caught most easily because they were the least active; their allocation to one treatment while the other 10 receive a second, on the assumption that they are a haphazard selection, will then produce a bias if the measurement eventually made is correlated with the activity of an animal. Moreover, an element of randomness in the allocation of subjects to treatments is strictly a prerequisite for the use of standard methods of statistical analysis, and any neglect of this will necessitate special explanations even if it does not invalidate the experiment.

3.3. Results and Statistical Analysis

In his experiment with 20 rats, Bacharach reported the results in Table 3.1. These may be used as an example of the reduction of measurements to counts for rapid analysis. An arbitrary dividing line can be taken, say 3,100 units, and the rats in each group classified as having vitamin A values above or below this level, in the manner shown in Table 3.2 (cf. Table 2.1).

The null hypothesis (§ 2.7) is that vitamin E deficiency did not affect the storage of vitamin A. The probability of

TABLE 3.1

VITAMIN A IN LIVERS OF TWENTY RATS
(International Units)

Normal* Diet	Diet Deficient in Vitamin E
3,950	2,650
3,800	3,350
3,450	2,450
3,350	2,650
3,700	2,650
3,900	3,150
3,800	2,900
3,050	1,700
2,650	1,700
2,000	2,500

* The "normal" diet in fact contained vitamin E far in excess of requirements.

TABLE 3.2

CLASSIFICATION OF RATS IN RESPECT
OF VITAMIN A IN LIVERS

Diet	Below 3,100 Units	Above 3,100 Units	Total
Normal.....	3	7	10
Deficient....	8	2	10
Total...	11	9	20

obtaining a difference between the proportions with high vitamin A as extreme as, or more extreme than, in Table 3.2 can then be found in the same way as for the clinical trial discussed in chapter ii: either samples from 20 cards may be

Measurements

made to correspond to hypothetical repetitions of the experiment, or an arithmetical procedure may be based upon the binomial distribution. The probability is 0.070. Despite the small numbers, the χ^2 calculation gives 3.23 and a probability of 0.072, which approximates well to the correct value. Thus the evidence presented in Table 3.2 does not justify the rejection of the null hypothesis. To most readers this must seem a strange conclusion from Table 3.1, the reason for which is undoubtedly the sacrifice of information about numerical magnitudes involved in the formation of Table 3.2.

Instead of comparing the difference between proportions of subjects in excess of 3,100 units (or any other arbitrary level) with an assessment of the variability in this difference that might be encountered if the null hypothesis were true, the preferred method of analysis is to compare the difference in the average amounts of vitamin A for the two groups of rats with an assessment of the variability to which the null hypothesis makes this quantity liable. The arithmetic mean or average vitamin A levels in the experiment were 3,365 and 2,570 for the normal and deficient diets, respectively. Any measure of the variability in individual rats on one diet must depend in some way upon the extent to which individual values differ from the mean for the treatment: statistical theory shows that, for most purposes, the best measure is based upon the sum of the squares of individual deviations from the mean. Denoting the individual values by x and their mean by $\bar{x}$, this sum of squares is the sum of all the values of $(x - \bar{x})^2$, written $S(x - \bar{x})^2$; for the normal diet

$$S(x - \bar{x})^2 = (3,950 - 3,365)^2 + (3,800 - 3,365)^2 + \ldots$$
$$+ (2,000 - 3,365)^2$$
$$= 3,600,250.$$

For simplicity and speed, especially when a calculating machine is used, a slightly different but equivalent formula is preferable (n is the number of observations in the group):

$$S(x^2) - \frac{\{S(x)\}^2}{n} = 3{,}950^2 + 3{,}800^2 + \ldots$$
$$+ 2{,}000^2 - \frac{(33{,}650)^2}{10}$$
$$= 3{,}600{,}250.$$

If $S(x - \bar{x})^2$ is divided by $(n - 1)$, one less than the number of observations, the result is the *variance*, and its square root is the *standard deviation* (S.D.).[1] The divisor of the sum of squares used in calculating the variance is known as the number of *degrees of freedom* (d.f.), because, in a sense that cannot be fully explained here, it represents the number of independent units of information on the variability inherent in the records. For information on these and other standard statistical terms, the reader should consult one of the books in Section A of the References. Here the standard deviation is 632: exact explanation of the meaning of this quantity is unnecessary for a book primarily concerned with experimental design and planning, and it will suffice here to state that, if a large number of similar rats were given this diet, the information available indicates that most (about 65 per cent) of their vitamin A values would be within 632 units of the mean and the great majority (90–95 per cent) within twice this range. The corresponding calculations for the deficient diet give

$$S(x - \bar{x})^2 = 2{,}606{,}000$$

and a standard deviation of 538.

The difference between the mean values of vitamin A in this experiment is 795 units. On the null hypothesis, repeti-

[1] More correctly, these are *estimates* from the 10 rats of what the variance and standard deviation would be in an indefinitely large set of similar and similarly treated rats.

tions of the experiment would be expected to give an assembly of values for this difference, some positive and some negative, that in their turn would have a mean zero. Such repetition of the experiment is obviously not practicable, and recourse must be had to a simple but very important piece of statistical theory. This is the theory that enables a measure of the variability in the difference in means for the hypothetical repetitions of the experiment to be formed from the variance of individual observations. A variance per observation (i.e., per rat) compounded from all the evidence is obtained by pooling the sums of squares of deviations and dividing by the total number of degrees of freedom: the result is s^2, where

$$s^2 = \frac{3{,}600{,}250 + 2{,}606{,}000}{9 + 9}$$

$$= 6{,}206{,}250 \div 18$$

$$= 344{,}800 \, .$$

Multiplication of this by the sum of the reciprocals of the numbers in the two groups,[2] here $(\frac{1}{10} + \frac{1}{10})$, gives the variance of the difference in means:

$$s^2(\tfrac{1}{10} + \tfrac{1}{10}) = 68{,}960 \, .$$

The reciprocals make appropriate allowance for a mean being less subject to variations than are single observations. The square root of this last variance, 263, is the *standard error* (S.E.) of the difference in means, and the probability that a single experimental value for the difference will differ from a value specified by hypothesis to any stated extent depends only on that standard error (and the degrees of freedom).

2. The statement in § 3.2 that the most precise comparison is obtained by using equal numbers in the two groups follows because, for a fixed total number of subjects, the sum of the reciprocals is least when the two numbers are equal.

Statistical Analysis

Table 3.3 is an extract from more extensive tables (e.g., Fisher and Yates, 1953) that simplify the evaluation of the probability. All that has to be done is to subtract the hypothetical difference from that found in the experiment and divide by the standard error: the result, generally denoted by t, is compared with the line of Table 3.3 for the number

TABLE 3.3

PROBABILITY LEVELS FOR t

d.f.	0.05	0.01
1	12.7	63.7
2	4.3	9.9
3	3.2	5.8
4	2.8	4.6
5	2.6	4.0
10	2.2	3.2
15	2.1	2.9
20	2.1	2.8
30	2.0	2.8
Very large	1.96	2.58

of degrees of freedom used in s^2. Here the null hypothesis specifies zero for the difference, and therefore

$$t = \frac{795 - 0}{263}$$

$$= 3.02 ,$$

which is slightly greater than the value for 18 d.f. in the column for a probability 0.01. Hence the probability is slightly less than 0.01, and on conventional standards the data have shown clearly significant evidence against the null hypothesis: we are obliged to conclude that the deficient diet does reduce the storage of vitamin A.

3.4. THE NORMAL DISTRIBUTION

Implicit in the analysis just described is the assumption that the variance in measurements among individuals

Measurements

treated alike conforms to what is known as the *Normal distribution*. The name is unfortunate and should not be taken as referring to "normality" of the animals in the colloquial sense: measurements that are not Normally distributed do not necessarily relate to abnormal circumstances, and, in order to emphasize the special use of the word "Normal," it is given a capital N throughout this book. Nevertheless, many biological measurements do manifest to a reasonable approximation the type of individual variation defined by the algebraic equation that comprises all Normal distributions, details of which can be found in most books on statistical analysis.

Although certainty that a particular series of measurements comes from a Normal distribution is rare, theoretical and empirical considerations justify the use of methods of analysis based upon it as a good approximation for many scientific problems. Tests of the adequacy of the approximation are beyond the scope of this book, and the experimenter must always be prepared to seek advice from a statistician in any case of doubt.

3.5. Homogeneity of Variance

A second assumption implicit in this and many other analyses is that the variance of individual measurements is unaffected by any experimental treatment, so that a composite or pooled estimate can be used for the whole experiment. Again much can be written about the justification for this, about tests for heterogeneity, and about the steps to be taken when the variance is not constant; again, fortunately, difficulties do not often arise in the simpler applications of statistical methods in biology, and discussion of them would be out of place here.

3.6. AN IMPROVEMENT IN DESIGN

An essential feature of the experiment as so far described was that the 10 rats on the deficient diet should be selected entirely at random from the 20 available. Many experimenters would consider that they could make a more precise and sensitive comparison between the diets by balancing the two groups in some way. Indeed, a common practice is to divide the animals into pairs such that the members of a pair are as alike as possible and then to assign one from each pair to each treatment. Provided that the selection from each pair for one treatment is made at random and independently of all other pairs (e.g., by spinning a fair coin once for each pair), this procedure is legitimate; in so far as the pairing succeeds in bringing together animals that are alike in the measurement studied except for effects of the treatment difference, it improves an experiment.

Any character that can be assessed before the experiment begins may be used as the basis of the pairing. The experimenter should try to use a character closely associated with the measurement eventually to be made, but his choice will be limited by convenience and practicability. For example, if the measurement to be studied on experimental animals were the weight of the heart, animals might be paired on the basis of likeness in initial body weight: to pair them on the basis of surface area would be laborious, and to pair them on initial heart weight impracticable! Quantitative characters, however, are not usually best employed in defining the pairs, since *covariance analysis* (§ 9.9) provides an alternative and better way of making allowance for their variations. Qualitative characters descriptive of the past or present environment (e.g., previous diet, position of cage, or season of year) or of the animals themselves and their genetic constitution (e.g., strain, litter, or sex) can be very valuable for this purpose.

In the rat experiment, pairs of litter-mates were in fact used, and the results in Table 3.1 are arranged so that pairs from one litter are on the same line. Although the last three litters had substantially less vitamin A than the others, the balancing should prevent this from affecting adversely the precision with which the effect of the deficient diet is estimated. The analyses in § 3.3, though correct if the experiment had been performed as described in § 3.2, are not appropriate to the paired design, but the preservation of an element of randomness in the allocation of rats to treatments insures that an analysis can be made. Once again a rapid test can easily be based upon the binomial distribution, exactly as in § 2.2. Of the 10 pairs of rats, 9 show a lower measurement on the deficient diet and 1 a higher. If the null hypothesis of no effect were true, positive and negative differences would be equally likely, and the probability of a deviation from equality as great as that observed could be found as in § 2.2. The result, 0.02, represents significant evidence that the deficiency reduces the vitamin A.

This test also has the flaw that it fails to use the information on actual numerical values. Again subject to certain assumptions of Normality, a better test can be made by forming the difference between control and deficient rats for each pair, estimating a variance of these differences, and comparing the mean difference with the corresponding value specified by the null hypothesis (zero) by use of the standard error of the mean. The mean difference is, of course, still 795 units, but the standard error is now only 167 units. The ratio

$$t = \frac{795 - 0}{167}$$
$$= 4.76$$

may again be referred to the t-distribution (Table 3.3), but the restriction on the randomization reduces the degrees of freedom from 18 to 9.[3] Although the value of t corresponding to any particular probability is greater than for 18 d.f., the great increase in t consequent upon the reduction in standard error more than compensates for this; the probability corresponding to 4.76 is approximately 0.001, thus leaving practically no room for doubt that the null hypothesis is false and that vitamin A storage is adversely affected by deficiency of vitamin E.

For this analysis to apply, the pairing must be an integral part of the structure of the experiment from the beginning. If in a fully randomized arrangement the measurements were grouped at random into pairs before analysis, no advantage would be gained (as, on an average, the variance would be unaffected), and degrees of freedom would be unnecessarily lost. On the other hand, if pairs were formed in accordance with some property of the measurements themselves (e.g., highest of the controls with highest of the deficient, second highest with second highest, and so on), the analysis using pairs would be biased. Chapter iii of Fisher's *The Design of Experiments* contains a more detailed discussion of these points.

3.7. ESTIMATION

Emphasis has been placed on the making of tests of significance, but more often the real purpose of an experiment is to estimate one or more treatment effects. Instead of the question "Does deficiency in vitamin E reduce the amount of vitamin A stored in the liver?", the experimenter may ask "By how much does deficiency in vitamin E affect the storage

---

3. In the completely randomized design, each treatment gave a sum of squares of deviations with 9 d.f., so leading to 18 d.f. for the pooled variance, whereas now the variance is formed only from the sum of squares of deviations of the 10 differences (one from each litter), which has 9 d.f.

Measurements

of vitamin A?" The second question is broader than the first and is of a more useful type: often the existence of an effect is a priori very likely, yet an experiment is needed for assessing its size.

Whether or not the observed difference between the means for two treatments is statistically significant, it is the best estimate of the average difference that would be obtained from unlimited repetitions of the experiment. The usual practice is to quote this estimate with its standard error:

$$795 \pm 263 \text{ units} \quad \text{and} \quad 795 \pm 167 \text{ units}$$

for the two analyses discussed previously. Only by a lucky accident will the observed difference be exactly the correct value, and the standard error gives a measure of the uncertainty inherent in quoting the estimate. If the standard error is multiplied by the value of t for the 0.05 probability level, the product is the width of an interval on either side of the observed mean within which the true mean is likely to lie, the word "likely" here corresponding to a *fiducial probability* or degree of faith of 0.95. The 0.05 values of t for the two analyses are 2.10 and 2.26, so that if the first analysis were appropriate, the *fiducial limits* would be 243 and 1,347 units, and if the second, they would be 418 and 1,172 units. The conclusion drawn from the experiment would be that the best estimate of the true mean difference was 795 units and that values outside the limits quoted were contradicted by the evidence.

Significance tests based upon classifying and counting measurements rather than using actual numerical values not only are less sensitive in the detection of small differences but also do not lead readily to the estimation of the magnitudes of effects and the assessment of fiducial limits for these.

3.8. PRECISION AND EFFICIENCY

By the *precision* of an experiment is meant, in general terms, the closeness with which it serves to estimate some quantity. Since the variance of a mean is obtained by dividing the variance per observation by the number of observations, the reciprocal of the variance per observation is an appropriate measure of precision. For example, if a change in conditions of experimentation on animals were to increase the variance per animal threefold, three times as many animals would be required on any treatment in order to estimate the mean for that treatment with the same variance as before; hence the inherent precision of the second experiment is only one-third that of the first. Similarly, the precision of estimation of a particular mean or of a difference between means is measured by the reciprocal of the variance of the quantity. The precision is also a measure of the sensitivity of an experiment when a significance test is used to examine the departure from a null hypothesis. Even when specified treatments are to be compared in an experiment of fixed size (e.g., with a limited number of animals), alternative designs may be available. The ratios betwen the precisions of the alternatives in respect of any quantity to be estimated then measure the *relative efficiencies* of the designs and indicate the extent to which the size of a less efficient design would need to be increased in order that it should give the same variance and standard error as a more efficient design. For example, in the vitamin experiment discussed previously, the standard error in § 3.3—263—is, in fact, an estimate of the S.E. that would be found if an experiment of the same size but without pairing (as in § 3.2) were conducted on a random selection of animals from the same source. Hence the efficiency of the paired design relative to the completely randomized, obtained as the inverse ratio of variances, is

$$E = \frac{263^2}{167^2}$$

$$= 2.48;$$

the pairing improved the experiment almost as much as an expansion of the completely randomized design from 20 rats to 50 ($2\frac{1}{2}$ times as many), and this gain is obtained in return for only a simple change in the conduct of the experiment.

3.9. A Further Complication

The experiment reported by Bacharach had still one more complicating feature. The first 5 pairs of rats were males and the others females, and the first 4 female pairs came from the same litters, respectively, as the first 4 male pairs. The introduction of the restriction that some pairs should be of one sex and some of the other has two merits: first, the two members of each pair are made more alike, and, secondly, the experiment now provides a test of whether the effect of vitamin E on vitamin A storage is the same for both sexes. The use of male and female pairs from the same litter is more debatable as an improvement, since conclusions on the existence and magnitude of any average effect of vitamin E are thereby based on an average of fewer litters.[4] The only compensation is that a supplementary comparison between the vitamin E effect on males and that on females may now be more precise, because any intralitter variation is eliminated. Rather too much seems to be attempted within a small experiment, but detailed consideration is beyond the scope of the present discussion.

4. The statistical analysis proposed in § 3.6 needs modification to take account of these changes in design and to examine new questions.

CHAPTER IV

Randomized Blocks and Latin Squares

4.1. Agricultural Research and Experimental Design

The first great stimulus to the development of the theory and practice of experimental design came from agricultural research. R. A. Fisher's recognition that current practices in field plot trials failed to produce unambiguous conclusions led him, from about 1923 onward, to examine the principles underlying scientific experimentation and to evolve new techniques of design. Not only was it necessary to devise procedures that would permit the drawing of valid inferences from experimental results, but these inferences had to be freed as far as possible from the obscuring effect of the variability inherent in the material and the nature of the observations. Not only was randomization needed in order to remove bias, and replication in order that valid estimates of standard errors might be derived, but the labor of performing experiments and the number of questions requiring investigation were so great as to make imperative techniques that should use most effectively the materials and effort employed and should give results of high precision. To Fisher belongs a great part of the credit for stating and solving these problems and so creating a new branch of science from which experimentation in many fields of research has since benefited.

Although this science of experimental design is today used

widely, in biology and elsewhere, the standard nomenclature retains evidence of its agricultural origin. The words taken over from agricultural research often help the reader to visualize a problem: they must never be thought to limit the application of the methods.

4.2. Experimental Units

In field experiments the ultimate experimental unit that is differentiated for the purpose of receiving a treatment (a fertilizer, a method of cultivation, a seed rate, a particular date of sowing, etc.) is the *plot*, a small area of land with dimensions chosen by the experimenter. The word is now used generally for the ultimate experimental unit, with the understanding that in particular applications of a design the plot may be something entirely different from an area of agricultural land. In the vitamin experiment of Table 3.1, individual rats play the part of plots; in other circumstances, the plot may be a hospital patient, a single leaf on a growing plant, a piece of animal tissue, a particular site of injection on the body of an animal, or even a group of animals in one cage treated as a unit for the purposes of the experiment.

4.3. Experiments on Several Treatments

In chapter iii, the problem of designing an experiment for comparing two alternative treatments has been considered in detail. Often an investigator wishes to compare several treatments, yet to plan a separate experiment for every pair would be extravagant. Indeed, even if such experiments were completed, the results would often be far from satisfactory because comparisons were not all made under the same conditions or because an essential feature of the investigation was to examine the interactions of various combinations of treatments. The principles of chapter iii, however, can be applied to simultaneous trial of any number of treatments. New difficulties in the conduct of an experiment may be raised by introducing many treatments, and an important duty for the statistician is to find ways of surmounting these without seriously impairing precision.

4.4. REPLICATION

Whatever the units to which treatments are to be applied, two or more plots must be allocated to each treatment, in order that account may be taken of individual variations between units treated alike. For the vitamin experiment of Table 3.1, if only one rat had been allocated to each of the two treatments, there would have been no way of judging whether an observed difference was the effect of treatment or was entirely due to chance: in fact, tenfold *replication* of each treatment was adopted. The need for replication does not mean that every combination of treatments must always be replicated on two or more plots (see §§ 6.8, 6.9).

4.5. RANDOMIZATION

The second essential feature of a good experiment is that of *randomization*. Arguments relating to this have been presented at length in §§ 2.9 and 3.2 and need not be repeated. If bias in the estimation of treatment differences and bias in the assessment of standard errors are to be avoided, the experimental units must be allotted at random to the treatments. This randomization need not be complete: it may be subjected to certain restrictions, provided that due allowance is made for them in the subsequent statistical analysis (§ 3.6). Neither haphazard nor deliberate selection is a permissible alternative to the strict objectivity of randomization. Experience has shown that an experimenter who adopts an arrangement that he considers "effectively random," without having used a recognized randomization technique, runs a grave risk of bias. Occasionally, practical difficulties make departure from true randomness inevitable: the imaginative statistician can then almost always think of ways in which bias might enter, statistical analysis can do practically nothing to indicate whether such a bias is present, and the experimenter can assert conclusions about the treatments tested only in so far as *he* is prepared to take the responsibility for asserting that the bias is nonexistent or trivial (cf. § 2.8).

In statistical contexts, randomness always implies selection between the permitted alternatives by a process equivalent to a perfectly fair lottery. In practice, it would suffice if experimenters were to draw lots with the aid of carefully prepared sets of numbered cards, but they can be saved this trouble by using *tables of random numbers*, those given by Fisher and Yates (1953) being the most readily accessible. These authors, Cochran and Cox (1950), and Quenouille (1953) have also published sequences that enable random orders for various numbers of entities to be written down directly.

Throughout this book, strict randomization will be assumed in respect of every design discussed. For example, in all experiments arranged in blocks (§ 4.7), the treatments that occur in a block are to be assigned at random to the plots. When plots (e.g., animals) are to be treated in a time sequence (e.g., § 7.5), each must be randomly selected from the population available. The safest rule for the experimenter is to make all the randomizations he can within the constraints of the definition of his design: *when in doubt, randomize.* Consultation with a statistician will help to discover whether any of these are unnecessary, whether any can be omitted without appreciable risk, and what are the major risks associated with omission of others.

It cannot be too strongly emphasized that randomization

is an integral part of the specification of a design, falling within principle iii of § 1.2. For example, the design shown in Plan 4.2 is a Latin square only in so far as the allocation of treatments was selected at random from the set of possible arrangements having the same restrictions on rows and columns. Exactly the same order of treatments on leaf sizes might have occurred in a randomized block design with the five plants as blocks. Consequently, inspection of Plan 4.2 does not suffice to identify the design, unless the imposed constraints and the rules of randomization are stated or implied. This book follows the generally accepted convention that, when an experimental plan is presented, the proper randomizations either have been performed (if a completed experiment is being described) or are to be performed (if an example of a type of design for future use is under discussion).

4.6. Completely Randomized Design

The experimenter who wishes to compare several treatments simultaneously faces essentially the same problem as that of § 3.2. His resources limit the total number of plots that can be used, and he must plan to make comparisons with maximum precision. Although he can increase the precision for the difference between one pair of treatments by allocating more plots to them, if all treatments are of equal interest the best procedure is to have equal numbers of plots of each.

The obvious generalization of the scheme of experimentation described in § 3.2 is the *completely randomized design*, in which the appropriate number of plots for each treatment is selected entirely at random from the total number available. For example, if the growth of four strains of bacteria were to be compared, the plot might be a single inoculated plate on which some assessment of growth (area or number of colonies) was to be made. If the total number of plates is limited, they should be divided at random into four equal groups to which the strains will be allocated at random.

The total number of plots is here assumed to be a multiple of the number of treatments. If not, it can be made so by discarding some plots or, since the conditions are seldom absolutely rigid, by adding a few. For the completely randomized design, the exact number of plots can be used by allowing some treatments one plot more than others, but for other designs this is rarely desirable.

The statistical analysis of completely randomized experiments has no difficulties for those familiar with other analyses described briefly later, but it will not be discussed here.

4.7. BLOCKS

Completely randomized experiments would often have much larger variances and standard errors than can be attained by quite simple modifications. The principle is that of the paired experiment in § 3.6, namely, balancing the treatments in respect of other characteristics (especially qualitative) of the plots. Groups of plots that share some property are made up in advance of the experiment (usually with equal numbers of plots per group), and members of a group are then assigned to different treatments at random.

These groups are termed *blocks*, another word from field plot trials where the device of balancing treatments over compact blocks of adjacent plots is used for the control and elimination of soil heterogeneity or other positional effects; each block there consists of plots in which soil fertility and other factors influencing plant growth, apart from the applied treatments, may reasonably be expected to be more homogeneous than over the whole experimental area. In other branches of research, a block may be a single litter of animals, a set of blood or serum samples obtained from one animal, a location in an incubator, a set of leaves on one plant, a series of determinations made on one day or by one man, or a set of inocula on one agar plate destined to receive doses of different antibiotic preparations. Any property of the plots that can be determined before an experiment begins can form the basis of a grouping into blocks: the judgment and experience of both the experimenter and the statistician are called into play in choosing properties easy to work with, yet likely to be so associated with the final measurement that balancing in respect of them can substantially reduce variation.

4.8. RANDOMIZED BLOCKS

The most valuable of all experimental designs, the most frequently used, and, except for the completely randomized, the simplest in construction and statistical analysis is the *randomized block design*. This is a natural extension of the randomized pairs described in § 3.6. The blocks are formed in such a way that each contains as many plots as there are treatments to be tested, and one plot from each is randomly selected for each treatment. The scheme is most readily understood by visualizing a field plan for an agricultural experiment, say for four treatments (A, B, C, D) in six blocks of four plots. The arrangement on the field might be as shown in Plan 4.1. The results would be recorded in a table of four columns (for the four treatments) and six rows (for the six blocks), a systematic order for ease of totaling and analysis, but randomization within each block on the field is essential.

This design is typical of many used in different branches of research. In animal experiments, litters are frequently used as blocks, one animal from each litter being assigned to each drug or diet or other treatment under test, in order that every difference between treatments shall be estimated independently of interlitter variation. Wadley (1948) reported the use of single cows as blocks in a comparison between three doses of each of two tuberculins: injections were made at fourteen sites on a cow, each dose at all sites, so that the "plot" consisted of an assembly of fourteen injection points for which the mean skin thickness was measured. The whole scheme was then replicated over five cows, thus giving 5 blocks of 6 treatments.

PLAN 4.1*

FIELD PLAN FOR A RANDOMIZED BLOCK EXPERIMENT

I		II		III	
A	D	B	C	C	A
C	B	A	D	D	B
C	D	B	D	C	D
B	A	A	C	A	B
IV		V		VI	

* Roman numerals denote the blocks, bounded by full lines; broken lines separate the plots.

Randomized blocks are also frequently wanted in tests of technique. Biggs and Macmillan (1948) wished to compare five doctors in the counting of red blood cells. To have made repeated tests always with the same apparatus would have left a danger that differences were peculiar to that apparatus. Instead, ten different pipettes and counting chambers were used, each doctor making one count with each. Here the blocks—the different pieces of apparatus—were used to give a broader basis for any inferences that might be drawn and also to supply information about differences between pipettes. Table 4.1 records fifty counts, all on the same sample of blood.

The statistical analysis of the experiment involves par-

Randomized Blocks

titioning the variation between all the observations into a component representing differences between pipettes, another representing differences between doctors, and a third from which the residual variance or *error* can be assessed. Table 4.2 shows the *analysis of variance* calculated from

TABLE 4.1

NUMBERS OF RED CELLS COUNTED BY FIVE DOCTORS

Doctor	Pipette and Counting Chamber									
	I	II	III	IV	V	VI	VII	VIII	IX	X
A	427	372	418	440	349	484	430	416	449	464
B	434	420	385	472	415	420	415	396	439	424
C	480	421	473	496	474	411	472	423	502	488
D	451	369	500	464	444	410	422	396	459	471
E	462	453	450	520	489	409	508	347	440	391

TABLE 4.2

ANALYSIS OF VARIANCE FOR TABLE 4.1

Adjustment for Mean		9,648,346	
Source of Variation	d.f.	Sum of Squares	Mean Square
Pipettes	9	26,721	2,969
Doctors	4	11,750	2,938
Error	36	42,327	1,176
Total	49	80,798	

Table 4.1; the method of calculation is explained more fully for a different design in § 4.11, and the reader should try to reproduce Table 4.2 after he has studied Table 4.5. This analysis, the most important single analytical technique in the biometric application of statistics, is explained in standard textbooks. Here note only that the mean square

for "doctors" can be compared with that for error in a test of significance; although the evidence of this experiment is thereby shown not to reject the null hypothesis (§ 2.7) that the five doctors, on an average, obtain equal counts, the test criterion almost reaches the 0.05 probability level (§ 4.11) and suggests that further study might disclose real differences. (A similar test with the "pipettes" mean square shows significant evidence of differences between pipettes, some consistently tending to give high counts and some to give low.) The analysis may seem entirely different from that used

TABLE 4.3

MEAN COUNTS FROM TABLE 4.1

\multicolumn{5}{c	}{Doctor}
A	B	C	D	E
424.9	422.0	464.0	438.6	446.9

Standard error: ±10.8

in § 3.6, but in reality the t-test there described is equivalent to an analysis of variance with only two treatments. In Table 4.2 the mean square for error is the variance per observation. The standard error of the mean count for each doctor is obtained by dividing the variance by the number of replications (10) and taking the square root; Table 4.3 summarizes these means.

Mather *et al.* (1947) give another example of the use of randomized blocks in a study of technique. The plasma volume in man may be estimated by injecting a known quantity of the dye Evans Blue into the circulatory system and measuring its concentration in a sample taken after complete mixing. In a study of the effect of length of time between injection and sampling on the concentration, six different times ranging from 15 to 90 minutes were to be studied. Although all samples could have been taken from one man, a wider basis for inference was wanted. Hence samples at each time were taken from each of five subjects. A slight modification in design was that on every occasion duplicate determinations of dye concentration were made (i.e., 60 observations in all, instead of 30), so that the importance of any variation in the time effect from subject to subject could be assessed against a measure of the variation from sample to sample in one man at one time.

The importance of the randomized block design lies in its great adaptability to widely different situations. A thorough understanding is essential to all who want to appreciate the character and use of other designs.

4.9. COUNTS AND MEASUREMENTS

Chapters ii and iii have emphasized strongly the contrast between counts and measurements in respect of the appropriate methods of statistical analysis, although (§§ 3.3, 3.6) rapid statistical tests on measurements are sometimes made by reduction to counts. Tables 4.1 and 4.2 exemplify the reverse procedure, a method of analysis developed for measurements on a continuous scale being applied to the necessarily discrete counts of red blood cells. This can always be done for an experiment in which comparable replicate counts are made under a number of different treatments, although the standard tests of significance for the analysis of variance table may be untrustworthy if the counts are small or excessively variable. When the counts are fairly large and all of much the same order of magnitude, as in Table 4.1, discontinuities of scale can be ignored, and other objections to the analysis of variance become of little account. Moreover, any nonindependence of the individuals counted, such as a tendency for "clumping" (groups occurring in close association) or for repulsion and excessively regular distribution, destroys the possibility of making use of theoretical probability distributions of counts (e.g., the Poisson distribution); tendencies of this kind are often found in counts of cells or, as another illustration, in insect infestations of plants or animals.

4.10. Latin Squares

As explained in § 4.7, blocks are usually chosen with a view to eliminating unwanted variation and increasing the precision of comparisons, although examples have been given in § 4.8 of their use to broaden the basis of inference. In both contexts, situations arise in which the experimenter has in mind two different types of grouping as a basis for his blocks and either can see no reason for ignoring one or suspects that each would be valuable. He may therefore wish to employ two block systems simultaneously. With suitable attention to randomization, this can be done; but the statistical analysis is excessively laborious unless the two block systems are related to each other and to the treatments in some symmetrical manner.

The simplest and most important design of this category is the *Latin square*, which takes its name from a form of mathematical puzzle that was studied many years before its use as a plan of experiment. The block systems are such that each block of either contains one plot from each member of the other; the two systems are generally distinguished as *rows* and *columns*. Moreover, each treatment occurs once in each row and once in each column. Thus the design can be used only if the number of treatments is the same as the number of plots per row and the number per column.

Cox and Cochran (1946) described an experiment for the

comparison of five virus inoculations of plants. The plot was a single leaf, and the two block systems were plants and leaf sizes. Five plants were taken, and five leaves on each plant: the design is shown as Plan 4.2, in which the columns were the plants and the rows were the five largest leaves, the five second largest leaves, and so on. The treatments, represented by letters, have been allocated in such a way that one leaf of each plant has each treatment and, of the five leaves receiving a particular treatment, one is the largest on its plant, one is the second largest, and so on.

PLAN 4.2

SCHEME FOR A PLANT VIRUS EXPERIMENT

Size of Leaf	Plant No.				
	I	II	III	IV	V
1	A	C	B	E	D
2	E	D	C	A	B
3	D	A	E	B	C
4	C	B	A	D	E
5	B	E	D	C	A

Latin squares are extensively used in agricultural trials in order to eliminate fertility trends in two directions simultaneously. An arrangement such as that in Plan 4.2 is then a physical reality on the ground: the plots lie in a square formation of rows and columns, although, of course, the plots themselves need not be square. In other fields of research, the square may be a logical rather than a physical relationship. Emmens (1948, § 6.5) gives results of an experiment on the thyroid weights of guinea pigs that received five different doses of thyrotrophin. Animals of five strains were kept in five cages with one from each strain per cage, and a Latin square determined the allocation of doses to strains and cages.

Harrison *et al.* (1951) have used squares as large as 12 × 12 in studies of the effect of changes in pH, and of the addition of potassium cyanide to the vitamin samples, on the growth of *Escherichia coli* supplied with different doses of vitamin B_{12}, the square permitting the elimination of positional effects on a large agar plate.

A Latin square for use should ideally be selected at random from all possible squares of the same size, but there are practical difficulties because for the larger squares the total numbers of possibilities are very large. The totals are given in the accompanying tabulation:

Size of Square	No. of Different Squares
2×2	2
3×3	12
4×4	576
5×5	161,280
6×6	812,851,200
7×7	61,479,419,904,000

No simple formula exists, and the totals for larger squares are not known. Fairly rapid procedures for the selection of a random square up to 7 × 7 have been devised (Fisher and Yates, 1953; Kitagawa and Mitome, 1953). From any Latin square, a new one can be constructed by interchanging two or more rows (keeping the order within the rows fixed), by interchanging two or more columns, or by interchanging the positions of two or more of the letters representing treatments. In practice, for the larger squares, any particular square can be taken as the basis of one for use, provided that first the rows are rearranged in random order (without altering order within a row), secondly the columns are rearranged in random order (without altering order within a column), and thirdly the letters are assigned in random order to the experimental treatments.

4.11. Statistical Analysis of a Latin Square

In a study of the effect of site of injection on the size of bleb produced in rabbits by testicular diffusing factor, Bacharach *et al.* (1940) used six rabbits and injected a standard dose at six sites on each: A, B, C near the vertebrae and D, E, F laterally. Fearing that bleb size might also be influenced by the order in which the six sites on a rabbit were injected,

TABLE 4.4

BLEB AREAS (SQ. CM.) AFTER INJECTION
OF TESTICULAR DIFFUSING FACTOR

Animal	Order of Injection						Total
	1	2	3	4	5	6	
I......	B 7.5	C 6.7	A 7.9	D 6.1	F 7.3	E 6.9	42.4
II.....	E 8.5	D 8.2	B 8.1	C 9.9	A 8.7	F 8.3	51.7
III....	C 7.3	F 7.3	E 6.8	A 7.4	B 6.0	D 7.7	42.5
IV.....	A 7.4	E 7.7	C 6.4	F 5.8	D 7.1	B 6.4	40.8
V......	F 6.4	B 6.2	D 8.1	E 8.5	C 6.4	A 7.1	42.7
VI.....	D 5.9	A 8.2	F 7.7	B 7.5	E 8.5	C 7.3	45.1
Total..	43.0	44.3	45.0	45.2	44.0	43.7	265.2

SITE TOTALS

A	B	C	D	E	F
46.7	41.7	44.0	43.1	46.9	42.8

they controlled both order and animal differences by the Latin square in Table 4.4. The table also shows the areas of blebs (sq. cm.) 20 minutes after injection.

A brief explanation of the computations required for the analysis of variance may be of interest as typifying the standard process for separating the sum of squares of the deviations of all observations from the general mean into components relating to different sources of variation. The first step is to form the totals shown in Table 4.4, by animals (rows), order of injection (columns), and sites of injection (letters), checking that each set of totals adds to the grand total, 265.2.

Table 4.5 may now be constructed in eight steps:

(i) Analyze the total of 35 d.f., one less than the total number of observations, into 5 d.f. for differences between the six animals, similarly 5 d.f. for order of injection, 5 d.f. for sites, and the remainder for error.

TABLE 4.5
ANALYSIS OF VARIANCE FOR TABLE 4.4

Source of Variation	d.f.	Adjustment for Mean	1,953.64
		Sum of Squares	Mean Square
Animals	5	12.83	2.566
Order	5	0.56	0.112
Sites	5	3.83	0.766
Error	20	13.14	0.657
Total	35	30.36	

(ii) Calculate the *adjustment for the mean* needed in forming the various sums of squares (cf. § 3.3):

$$\{S(x)\}^2 \div n = (265.2)^2 \div 36$$
$$= 1,953.64 \,.$$

(iii) The sum of squares of all deviations is (reading down columns in Table 4.4)

$$7.5^2 + 8.5^2 + 7.3^2 + \ldots + 7.1^2 + 7.3^2 - 1,953.64 = 30.36 \,.$$

(iv) The sum of squares for differences between animals is

$$(42.4^2 + 51.7^2 + \ldots + 45.1^2 - 6 \times 1,953.64) \div 6 = 12.83 \,.$$

Analysis of a Latin Square

(v) Similarly, for order of injection,

$(43.0^2 + 44.3^2 + \ldots + 43.7^2 - 6 \times 1{,}953.64) \div 6 = 0.56$.

(vi) Similarly, for sites of injection,

$(46.7^2 + 41.7^2 + \ldots + 42.8^2 - 6 \times 1{,}953.64) \div 6 = 3.83$.

(vii) Subtract items iv, v, and vi from iii, the result being the error sum of squares.

(viii) Divide each of the first four entries in the sum-of-squares column by the corresponding number of degrees of freedom to give the column of mean squares.

In items iv, v, and vi, the multiplier and divisor 6 enters because the relevant totals (animals, order, sites, respectively) all consist of six of the original measurements and *not* because there are six totals in each category. The distinction is unimportant here but is important for Table 4.2, which the reader should now have no difficulty in computing by similar steps. Comparison of the mean squares leads to tests of significance.[1] For example, if a null hypothesis of no real difference in animals in respect of potential bleb size is true, the ratio of mean squares for animals and error,

$$F = \frac{2.566}{0.657}$$

$$= 3.91,$$

with 5 and 20 d.f., has a probability of little more than 0.01 of being attained: hence this hypothesis can be dismissed, and an association between bleb size and animal differences is established. On the other hand, the ratio of mean squares for sites and error,

$$F = \frac{0.766}{0.657}$$

$$= 1.17.$$

1. By reference to standard tables, such as Fisher and Yates's (1953) Table V.

is not statistically significant. Table 4.6 shows mean bleb areas for the six sites, with their standard error. No strong evidence for any real effects of site differences appears, even the main contrast of median and lateral sites apparently having little effect. Evidence for association of bleb area with order of injection is also not statistically significant.

TABLE 4.6

MEAN BLEB AREAS (SQ. CM.) FROM TABLE 4.3

\multicolumn{6}{c}{Site of Injections}
\multicolumn{3}{c}{Median}	\multicolumn{3}{c}{Lateral}
A	B	C	D	E	F
7.78	6.95	7.33	7.18	7.82	7.13

Standard error: ±0.331

4.12. Orthogonality

The type of balance of treatment and block constraints achieved in the randomized block and Latin square designs is known as *orthogonality* and is immensely important in the theory and application of experimental design. In any design, two classifications (such as treatments and blocks) are said to be orthogonal if the difference between every pair of means for one classification (e.g., treatments) involves taking as many plots negatively as positively from each member of the other classification. This property is necessarily reciprocal, in that a difference between a pair of means for the second classification is similarly balanced for the first. In the randomized block design of Table 4.1, for example, the difference between mean counts for any two doctors involves one positive and one negative "plot" from each of the ten blocks. In the Latin square of Table 4.4, treatments are orthogonal

Orthogonality

with animals and also with order, and animals are orthogonal with order. The analysis of the total sum of squares of deviations into independent components, illustrated in Tables 4.2 and 4.5, is made possible by orthogonality.

4.13. GRAECO-LATIN SQUARES

Randomized blocks impose one constraint on the allocation of treatment to plots; Latin squares impose two. Designs can be constructed in which three or more are imposed simultaneously. For example, in the situation that gave rise to Plan 4.2, the experimenter might have wished to inoculate

PLAN 4.3

MODIFIED DESIGN FOR A PLANT VIRUS EXPERIMENT

PLANT NO.	SIZE OF LEAF				
	i	ii	iii	iv	v
I	Aγ*	Eβ	Dϵ	Cα	Bδ
II	Cϵ	Dδ	Aβ	Bγ	Eα
III	Bβ	Aα	Cδ	Eϵ	Dγ
IV	Eδ	Cγ	Bα	Dβ	Aϵ
V	Dα	Bϵ	Eγ	Aδ	Cβ

* Greek letters denote occasions.

on five different days and to balance occasions over treatments, plants, and leaf sizes. This cannot be arranged with the Latin square in Plan 4.2, but a few changes make it possible: in Plan 4.3, the Greek letters are so located that each occurs once for each plant, once for each leaf size, and once with each inoculation. The resulting design is known as a *Graeco-Latin square*. Each of the four classifications is orthogonal with the other three, and the statistical analysis is a simple extension of that for the Latin square. The idea can be generalized so as to include more orthogonal classifications, up to a maximum of $(k + 1)$ for a $(k \times k)$ square (cf. § 5.5).

Graeco-Latin squares are far less numerous than simple Latin squares, and, if the Latin square is first chosen, to superimpose a Greek square may be difficult or even impossible; it is usually preferable to start from a known Graeco-Latin square and obtain one for use in an experiment by permutations of rows, columns, and letters (§ 4.10). For 2×2, 6×6, and 10×10 arrangements, no Graeco-Latin squares exist, but, except for the trivial 2×2, other slightly less elaborate orthogonal schemes can be devised. The 6×6 Latin square in Plan 4.4 was used in a study of the histaminase activity of sera from pregnant women. Tests with histamine-histidine mixtures in six different proportions (A, B, ..., F) were to be made on sera from six subjects, and, since the order in which six tubes were poured from a sample of serum might influence the results, a Latin square was used to determine the allocation of the six mixtures to the combinations of subject and order in which a tube was poured. Suppose that a further balancing were required with respect to some other factor (such as the use of different instruments or operators in reading the results of the tests); this would not be possible if the extra factor were at six levels, except by associating it completely with subjects or order (so that, for example, all tests for one subject were read by the same operator). Plan 4.4, however, shows how a new factor at two levels only (α, β) can be simultaneously balanced over subjects, order, and histamine-histidine treatment: if two operators were to share the work, each could do 18 tests consisting of 3 from each level of treatment, 3 from each patient, and 3 from each position in the pouring order. Other orthogonal partitions are possible, at least for some 6×6 Latin squares, such as balancing in respect of a new factor at three levels.

4.14. SETS OF LATIN SQUARES

A single small Latin square may not provide adequate replication and so may not estimate differences with sufficient precision. Several squares with the same treatments can be used and included in a comprehensive analysis. The squares may be entirely independent or may have their rows (or columns) coinciding, with slight consequential differences in the form of the analysis of variance. For example, in the ex-

PLAN 4.4

LATIN SQUARE DESIGN FOR EXPERIMENT ON SERA FROM PREGNANT WOMEN, WITH AN ORTHOGONAL PARTITION

Subject	Order in Which Tube Was Poured					
	1	2	3	4	5	6
I	Ba	Eβ	Da	Fβ	Ca	Aβ
II	Fa	Ba	Cβ	Aβ	Dβ	Ea
III	Aa	Ca	Fβ	Ba	Eβ	Dβ
IV	Cβ	Aβ	Ea	Da	Fa	Bβ
V	Dβ	Fβ	Bβ	Ea	Aa	Ca
VI	Eβ	Da	Aa	Cβ	Bβ	Fa

periment reported in Table 4.1 a possible modification would have been to have the doctors make counts on several samples of blood. Two sets of five samples might have been taken, the first being associated with pipettes I–V and the second with pipettes VI–X; doctors would then have been assigned to combinations of pipettes and samples with the aid of two 5 × 5 Latin squares. Alternatively, only five blood samples might have been used, so making the rows of the two squares coincide, as shown in Plan 4.5. When several squares are wanted in one experiment, they should be selected by entirely independent randomizations.

Cochran *et al.* (1941) have illustrated the value of Latin squares for experiments in which the units can receive several treatments in succession. For example, columns of a square can correspond to different animals, rows to a succession of dietary treatments; the comparison between treatments in respect of measurements (say of milk production) during the various periods is freed from interanimal variation. Of special importance is the possibility of using a balanced set of

PLAN 4.5

MODIFIED DESIGN FOR THE EXPERIMENT IN TABLE 4.1

BLOOD SAMPLE	PIPETTE AND COUNTING CHAMBER									
	I	II	III	IV	V	VI	VII	VIII	IX	X
1......	B	D	E	A	C	A	E	C	B	D
2......	A	C	D	E	B	B	C	D	E	A
3......	D	A	C	B	E	D	A	B	C	E
4......	C	E	B	D	A	E	B	A	D	C
5......	E	B	A	C	D	C	D	E	A	B

PLAN 4.6

DESIGN FOR AN EXPERIMENT ON ANIMAL NUTRITION

PERIOD	ANIMAL No.											
	1	2	3	4	5	6	7	8	9	10	11	12
I.....	D	C	A	B	A	D	C	B	A	B	C	D
II....	B	A	C	D	B	C	D	A	D	C	B	A
III....	A	B	D	C	C	B	A	D	B	A	D	C
IV....	C	D	B	A	D	A	B	C	C	D	A	B

squares in such a way that each treatment in each period is preceded by every treatment on one or more animals; residual effects of treatments can then be estimated in order to improve the evaluation of the relative merits of the treatments. Plan 4.6 shows such a design for four treatments, using

Sets of Latin Squares

twelve animals in three 4 × 4 Latin squares. Others have extended and improved the usefulness of designs of this type, one important suggestion being the addition of an extra period in which the last row of each Latin square is repeated.

4.15. LATIN CUBES

The basic idea of a Latin square can be extended to patterns in three dimensions or more, but practical applications of Latin cubes and related designs are few.

CHAPTER V

Incomplete Block Designs

5.1. LIMITATIONS ON BLOCK SIZE

For randomized blocks or Latin squares, the number of plots per block (or per row and column) must equal the number of treatments. This may prove inconvenient or impracticable if the number of treatments is large: the purpose of a block arrangement is to make the precision of comparisons between treatments dependent only on inherent variability between plots of the same block, but its advantages are lost for blocks so large that their constituent plots are very heterogeneous.

In agricultural experiments the plots are small areas of crop, and blocks are designed for homogeneity in fertility and other inherent characteristics; with plots of ordinary size, blocks of as many as 16 or 20 plots may fail to control soil heterogeneity adequately, though, when plots are for special reasons very small, larger blocks can sometimes be used. With a Latin square, a smaller number of treatments is desirable, since rows or columns that are long narrow strips of land are less likely to be homogeneous than equal but more compact areas. For other purposes, block size may be more severely limited. When an animal experiment is to use littermate control, the smallest litter constitutes an upper limit to block size; if the experiment is restricted to animals of one sex, this upper limit may be as low as 2 or 3. In an experiment on virus inoculations for which plants form blocks with

Limitations on Block Size

leaves as plots, the number of usable leaves may be as low as 5 or even 3. In trials on human subjects, it may be possible to use subjects as blocks with successive tests of different treatments as plots, but the number of tests that individuals can be persuaded to undergo limits block size. If the nature of the experiment does not impose limits on block size, experience of similar research should be drawn upon to indicate what size is reasonable.

If a partial loss of orthogonality of treatment and block comparisons, with a consequent increase in the complexity of statistical analysis, is accepted, various types of *incomplete block design* can be devised; a high degree of symmetry can be retained so as to keep new difficulties to a minimum and to maximize the precision of comparisons.

5.2. AN EXPERIMENT ON SELF-ADMINISTERED ANALGESIA

Seward (1949) wished to compare a 1:1 mixture of nitrous oxide and air (A), a 3:1 mixture of nitrous oxide and oxygen (B), and a mixture of 0.5 per cent trichlorethylene and air (C) in self-administered analgesia for the relief of labor pains. Their efficiency was to be judged from the subjects' own statements, and, since no absolute scale of measurement was possible, each subject had to make at least two trials in order to be able to express a preference. The trials had to be made near the end of the first stage of labor, and a patient needed access to an analgesic for about half an hour in order to give it fair trial. Hence it was not practicable to have one patient test more than two of the mixtures. The scheme adopted used one of the simplest of incomplete block designs and illustrates how a well-designed experiment may yield clear conclusions without elaborate statistical analysis.

The experiment was based on 150 subjects in one hospital, each receiving two of the three mixtures and stating after-

Incomplete Block Designs

ward which was the more effective in relieving the pain of uterine contraction. Each of the three possible pairs of mixtures was assigned to 50 subjects; in order to balance residual effects or any tendency of the subjects to prefer, say, the latest method tried, irrespective of its analgesic effects, 25

PLAN 5.1

DESIGN FOR AN EXPERIMENT ON SELF-ADMINISTERED ANALGESIA

Period	Subjects' Nos.*					
	1–25	26–50	51–75	76–100	101–25	126–50
First ½ hour.....	A	B	A	C	B	C
Second ½ hour...	B	A	C	A	C	B

* The numbers attached to the subjects do not represent the sequence of cases. The criterion for inclusion in the experiment was a reasonable prospect of normal labor; on admission, such cases were assigned at random to the six groups, with the restriction that 25 be placed in each.

TABLE 5.1

RESULTS OF EXPERIMENT ACCORDING TO PLAN 5.1

Subjects Testing	Preferred Mixture			No Preference
	A	B	C	
A, B..........	0	49	....	1
A, C..........	0	....	45	5
B, C..........	....	14	12	24

had the mixtures in one order and the other 25 in the reverse (Plan 5.1). The results are summarized in Table 5.1. They show convincingly that the mixture of nitrous oxide and air was regarded as inferior and that subjects observed no consistent difference between the other two mixtures.

5.3. Balanced Incomplete Blocks

If the number of plots per block is less than the number of treatments to be tested, it is reasonable to require that every treatment be assigned to the same number of plots. A further condition that every pair of treatments shall occur equally often as "block-mates" insures that the standard error for a difference between two treatments is the same for every pair. Two simple examples of such *balanced incomplete block* designs will make the principle clear. One extreme is needed when blocks can consist of only two plots, as when monozygotic twins form the blocks or in tests of virus inoculations under conditions that permit a single leaf to be a block with different treatments on the two halves (Spencer and Price, 1943; Price, 1946), and all possible pairs of treatments must be used as blocks. If six treatments were to be tested, the blocks would consist of the 15 pairs

A, B;   A, C;   A, D;   A, E;   A, F;   B, C; . . . ; D, F;   E, F;

and an experiment would have 15 blocks or some multiple of 15. More generally, if v treatments are to be tested,

$$b = \tfrac{1}{2}v(v-1)$$

blocks (or some multiple of this) are needed. The other extreme is that of blocks one plot too small to accommodate all treatments. A balanced design is then obtained by using a number of blocks equal to the number of treatments and omitting each treatment in turn: with five treatments, the blocks would be

B, C, D, E;   A, C, D, E;   A, B, D, E;   A, B, C, E;   A, B, C, D.

Many balanced incomplete block schemes do not require blocks of every possible constitution. Moore and Bliss (1942) compared the toxicity to *Aphis rumicis* of six glycinonitrile compounds with that of a standard nicotine spray. Only

Incomplete Block Designs

three sprays could be tested on one day, since the tests required the use of several concentrations of a spray on different batches of aphids so that the median lethal concentration[1] could be estimated. The susceptibility of the aphids was expected to vary from day to day, and the plan adopted (Plan 5.2) was to use seven different blocks of three sprays on seven

PLAN 5.2

DESIGN FOR AN INSECTICIDE TOXICITY EXPERIMENT

Day	I	II	III	IV	V	VI	VII
Sprays	A, B, D	A, C, E	C, D, G	A, F, G	B, C, F	B, E, G	D, E, F

PLAN 5.3

DESIGN FOR AN EXPERIMENT ON THE PAIN OF PENICILLIN INJECTIONS*

Subject	Doses	Subject	Doses
I	A, B, C	VI	B, C, F
II	A, B, D	VII	B, D, E
III	A, C, E	VIII	B, E, F
IV	A, D, F	IX	C, D, E
V	A, E, F	X	C, D, F

* A further modification, adopted in order to balance the sites, was that the first dose shown for subjects I–X was injected at site 1, the second at site 2, and the third at site 3; twenty more subjects were then introduced, so that subjects XI–XX received the same triads of doses with the order of sites 2, 3, 1, and subjects XXI–XXX had the sites in the order 3, 1, 2. This change in fact made the design no longer simply of the balanced incomplete block type.

days. Each spray was tested three times in all, and every possible pair of sprays (21) occurred once as contemporaries.

Herwick *et al.* (1945) described an experiment on the relationship between dose of penicillin and the degree of pain produced at three different sites of injection. Six doses (A, B, ..., F) were assigned in threes to 10 subjects (I, II, ..., X), as shown in Plan 5.3. Every dose is repeated five times and

1. The concentration for which the average mortality is 50 per cent.

every pair of doses occurs in two blocks: C, E are block-mates for subjects III and IX.

A balanced incomplete block design may be described in terms of the number of units or plots per block (k), the number of treatments (v), the number of replicates or plots of each treatment (r), and the number of blocks (b). Obviously,

$$kb = vr,$$

since either is the total number of plots in the experiment. Moreover, the total number of plots in blocks containing a particular treatment is kr, and the definition of balanced incomplete blocks requires that the plots other than those of the particular treatment shall be equally divided between the remaining $(v - 1)$ treatments. Hence

$$\lambda = \frac{r(k-1)}{v-1}$$

must be a whole number. For many, but not for all, sets of numbers k, v, r, b satisfying those two conditions, balanced incomplete block designs exist (Fisher and Yates, 1953, Tables XVII–XIX).[2] For example, the reader may verify that 11 treatments can be arranged in blocks of 6 by taking A, B, D, E, F, J as the first block and writing the others as *cyclic permutations* of this: the second block is derived from the first by writing the next letters in alphabetic order to that set of 6 (i.e., B, C, E, F, G, K), and similar steps generate the remaining 9 blocks with the convention that K is followed by A in order to close the cycle. This design has 11 blocks and 6 replicates of each treatment, so that $\lambda = 3$. If the simplest arrangement for particular values of k and v does not give

[2]. General theory relating to the existence of designs is difficult. Two interesting conditions are that no balanced incomplete block design can have r smaller than k and that, if v is an even number, no design with $r = k$ exists unless $(k - \lambda)$ is a perfect square. Even the satisfying of these conditions, however, is no guarantee that a design can be constructed.

sufficient replication, it can be used several times over as part of one experiment (with independent randomizations), so that r, b, and λ are all increased by the same factor.

For an experiment, the letters used in specifying a balanced incomplete block design should be assigned at random to the treatments, and the treatments for a block should be assigned at random to the plots.

5.4. YOUDEN SQUARES

Youden square designs permit the use of two systems of blocks simultaneously (cf. § 4.10). These were first suggested by Youden (1937) for the investigation of inoculations of

PLAN 5.4

DESIGN FOR AN EXPERIMENT ON TOBACCO MOSAIC VIRUS

Position of Leaf	Plants							
	I	II	III	IV	V	VI	VII
Lower	D	A	G	F	C	B	E
Middle	A	C	D	G	B	E	F
Highest	B	E	C	A	F	G	D

plants with tobacco mosaic virus; they combine one set of complete blocks with one set of balanced incomplete blocks. Youden used plants as "columns" of his square, leaves as plots, and the relative position of leaves on the stem as "rows"; thus his experiment was similar to that of Plan 4.2, but with an incomplete replicate on each plant. Plan 5.4 shows a design for testing seven virus inocula, in which each treatment is tested once at each leaf position and the treatments assigned to different plants form an incomplete block scheme. Exactly the same design might have been used in the experiment of Plan 5.2 if Bliss and Moore had wished to balance the testing of their sprays over three times of day.

Youden Squares

Any balanced incomplete block design that has its number of blocks equal to its number of treatments can be arranged as a Youden square. The example in § 5.3 with $v = b = 11$ automatically appears in this form by writing the first block as the first column and completing each row with the full cycle of 11 letters, A, B, C, ..., K in alphabetic order, beginning with the letter in the first column and following K by A where necessary. Omission of one row (or column) from a Latin square gives a Youden square. Thus the design shown in Plan 5.1 is a set of simple Youden squares, 25 of type

```
A   C   B
B   A   C
```

formed by subjects 1–25, 76–125, and 25 of type

```
B   A   C
A   C   B
```

formed by the remainder. The experiment of which Plan 5.3 shows a part could not be arranged as a Youden square because v and b were unequal, but a generalization of the idea was achieved with 30 subjects by permuting the allocation of doses to sites.

Latin squares from which two or more rows have been removed or from which one row and one column have been removed may occasionally be useful because of limitations of experimental material. Yet other possibilities are the addition of extra rows or columns or the addition of a row and removal of a column. Sometimes a design conceived as a Latin square but lacking all plots of one or two treatments may be particularly suitable for an experiment. These designs are not Youden squares, although to some extent they are similar. The statistician needs to have in mind such variations on the theme, but their lesser symmetry reduces their practical value and increases the labor of statistical analysis.

Incomplete Block Designs

Before use, Youden square and related designs should be fully randomized in the same way as a Latin square (§ 4.10).

5.5. LATTICE DESIGNS

When many treatments must be tested in small blocks, balanced incomplete blocks may require an excessively large number of replications. If a further sacrifice of balance is accepted, *lattice designs* can be used. These are constructed by arranging the treatment symbols on a grid or lattice and constructing blocks from rows and columns. This is particularly useful for a number of treatments that is a perfect square; the case of 16 treatments provides an easily handled example, although the practical importance of the designs is greatest for larger numbers. If the treatments are written in random order into a 4 × 4 lattice, as

```
G   A   E   J
L   H   B   I
M   P   N   F
D   O   C   K
```

two types of block may be formed, one from rows and one from columns. These are listed as Blocks I–VIII of Plan 5.5, and an experiment of lattice design could consist of these alone. Not surprisingly, two treatments such as C and D that occur in the same block (IV) would be compared rather more precisely than two such as A and B that are never blockmates. If more than two complete replicates could be undertaken, one or both of these sets of four blocks could be repeated, but a better plan (because it comes nearer to balancing comparisons between treatments) is to introduce a third set of blocks consisting of groups of treatments orthogonal to rows and columns; further sets of blocks orthogonal to the first three can be added if the amount of replication to be undertaken permits this. The reader should verify that Blocks IX–XII in Plan 5.5 correspond to a Latin square

Lattice Designs

superimposed on the original 4 × 4 lattice: each of these four blocks contains one treatment from each row and one from each column. Similarly, Blocks XIII–XVI correspond to a Graeco-Latin square superimposed on the lattice. Blocks XVII–XX complete the possibilities of this kind of arrangement by providing one more orthogonal set of blocks; no larger number is possible, and, indeed, the 20 blocks give the particular form of balanced incomplete block design known as

PLAN 5.5

LATTICE DESIGN FOR COMPARING 16 TREATMENTS

Block   I: G, A, E, J   Block   V: G, L, M, D
Block   II: L, H, B, I   Block   VI: A, H, P, O
Block III: M, P, N, F   Block VII: E, B, N, C
Block   IV: D, O, C, K   Block VIII: J, I, F, K

Block   IX: G, H, N, K   Block XIII: G, O, B, F
Block   X: L, A, C, F   Block XIV: L, P, E, K
Block   XI: M, O, E, I   Block   XV: M, H, C, J
Block XII: D, P, B, J   Block XVI: D, A, N, I

Block XVII: G, P, C, I
Block XVIII: L, O, N, J
Block   XIX: M, A, B, K
Block   XX: D, H, E, F

a *balanced lattice*. An experiment could be based upon any two, three, or four of the five sets of four blocks, however, instead of on the fully balanced design. The order of treatments would be randomized independently in every block, exactly as for randomized complete blocks (§ 4.8).

Situations requiring two systems of blocks can be dealt with by making one set of blocks into rows and another into columns simultaneously. One 4 × 4 square of treatments (in fact, the original lattice) can be formed with Blocks I–IV of Plan 5.5 as rows and Blocks V–VIII as columns; and a second square could have Blocks IX–XII as rows and Blocks XIII–XVI as columns. If a third replicate were wanted, it could

Incomplete Block Designs

have Blocks XVII–XX as rows and Blocks I–IV as columns. These squares are easily written down, the second one being

```
G   K   H   N
F   L   C   A
O   E   M   I
B   P   J   D
```

Such a *lattice square* design is again an analogue of the Latin square. If every block of Plan 5.5 is used once as a row and once as a column, full balance according to the balanced incomplete block restrictions is achieved in rows and in columns; this is a *balanced lattice square* design. When the number of treatments ($v = k^2$) is the square of an odd number, balance can be achieved in $\frac{1}{2}(k + 1)$ squares by having each block of the balanced lattice system appear as a row *or* as a column; when k is even, balance requires $(k + 1)$ squares.

Other lattice designs can be formed from cubic arrays of treatment symbols. For example, 27 letters might be written in a $3 \times 3 \times 3$ cube and 9 blocks of 9 formed by plane sections in each of the three directions; alternatively, 27 blocks of 3 can be formed by lines in each direction. The principle can be extended to numbers of treatments that are higher powers of integers (e.g., $32 = 2^5$). Yet other designs, *rectangular lattices*, can be constructed for a number of treatments that is a product of two unequal integers, the most useful in practice being those of type 4×5, 5×6, 6×7, etc.

5.6. Partially Balanced Incomplete Blocks

Lattice designs fall within a wider category of *partially balanced incomplete blocks*, which generalize the requirements for balanced incomplete blocks at the cost of needing more laborious statistical analysis and no longer having the same variance for the difference between every pair of treatments. However, the combinations of v, k, and r that can be covered by balanced incomplete blocks are severely limited, and the lattices and other forms of partial balance extend the range of possibilities. Even then, not all the schemes that might seem to be wanted can be obtained without excessive replication or adoption of a design that has many different variances for treatment comparisons (§ 9.3).

5.7. Analysis of Incomplete Block Designs

The statistical analysis of incomplete block designs is much more laborious than that of randomized blocks and Latin squares, on account of the nonorthogonality of treatments and blocks. Essentially, the analysis consists in the solution of $(v + b - 1)$ linear equations as a preliminary to the analysis of variance. For the important classes of design, computing routines have been devised that achieve this as expeditiously as possible (Cochran and Cox, 1950; Fisher and Yates, 1953).

In incomplete block designs, information on differences between treatments is obtainable from comparisons between blocks as well as within blocks. For example, in Plan 5.2, intrablock estimation of the difference between treatments D and F can be based on a direct difference in Block VII and also on "chains" such as (D–A) from Block I plus (A–F) from Block IV, or (D–B) from Block I plus (B–F) from Block V, there being four such chains. In addition, Blocks (I + III) contain A, B, C, G, as well as D twice, and Blocks (IV + V) contain A, B, C, G, with F twice, so that half the difference between these totals is an estimate of the mean difference between D and F. Provided that the different types of block are allocated at random to their locations on the ground, or to whatever other properties of the experimental material are to define them, this interblock estimate can be combined with the intrablock, in order to use the entire information most effectively.

5.8. Use of Incomplete Block Designs

An incomplete block design may be adopted, as explained in § 5.1, either because blocks consisting of all treatments would be so large as to lower precision or because the nature of the experiment renders complete blocks impossible. The standard arithmetical analysis minimizes the computing labor, while insuring that the precision of treatment comparisons is, at worst, lower than that for complete blocks to only a trivial extent (by the utilization of interblock information: § 5.7) and, at best, substantially higher.

Nevertheless, incomplete block designs ought not to be chosen without careful thought. The experimenter should not be unnecessarily restrictive in his specification of the number of treatments to be tested, the number of plots per block, or the number of replicates of each treatment (§ 9.3). If the statistician is allowed a little freedom to vary these, he may be able to devise a much more satisfactory design. He will aim at balance, or near-balance, in order to avoid making some comparisons much less precisely than others, and a slight change in v or k may greatly affect this possibility. Moreover, the labor of statistical analysis is reduced if a design with a high degree of symmetry can be substituted for one with less. In an agricultural experiment that is to run for a year or more and is to consume much time and labor in its management, whether the statistical analysis occupies a skilled statistician for several days or a junior computer for one day may be of little moment. If essentially the same experimental design, at least in its statistical aspects, is to be used for a laboratory experiment on which all operations and measurements will be completed in an afternoon, this question assumes greater importance: the experimenter can now scarcely ignore the statistician's claim for consideration of minor changes in treatments and blocks that would reduce the labor.

5.9. OTHER DESIGNS

The incomplete block designs described here are not the only useful schemes for dealing with large numbers of treatments. Their attempts at balance and the ensuing complexity of structure may be disadvantageous in some circumstances. Often a large number of treatments will contain one or two whose performance is very different from the rest: their failure or extraordinary success may make it necessary to exclude them from the main statistical analysis and to present their results separately. Although statistical analysis is still possible, the labor of it may be vastly increased by the consequent loss of symmetry. Accidental losses of observations, which, undesirable as they are, occur sometimes in large experiments through external circumstances damaging to a particular plot or to the observations upon it, have similar consequences.

Designs in which all treatments are divided into groups in one way only, with one or two control treatments included in every group, avoid some of these disadvantages. Randomized blocks of every group are included in one experiment, and two treatments that are not block-mates can be compared in terms of the extent to which each differs from the control. Another possibility, particularly appropriate when blocks correspond to physical location, is to arrange all treatments in randomized blocks but to include in each block a systematic pattern of a control treatment; for every plot, an index of its expected performance on the control treatment can be constructed as an average of neighboring controls, and a covariance analysis (§ 9.9) between the measurements actually made and this index should go far to reduce the variance in large blocks. Exclusion of certain treatments, blocks, or plots from the analysis of experiments of either of these types is relatively simple.

CHAPTER VI

Factorial Experiments

6.1. Factorial Design in Agricultural Research

In the endeavor to improve the logical foundations of scientific experimentation, *factorial design* has proved one of the most fruitful developments. To those familiar with modern agricultural research, it may now be difficult to realize that Fisher (1926) should ever have needed to write: "No aphorism is more frequently repeated in connection with field trials, than that we must ask Nature few questions, or, ideally, one question, at a time. The writer is convinced that this view is wholly mistaken. Nature, he suggests, will best respond to a logical and carefully thought out questionnaire; indeed, if we ask her a single question she will often refuse to answer until some other topic has been discussed." The factors affecting the growth and yield of a crop—manuring, seed rate, methods of cultivation, dates at which various operations are performed, and so on—are many, and the effect of any one may be dependent upon conditions in respect of others. Conclusions from an experiment to determine the optimal amount of phosphatic fertilizer to apply to a crop would become useless if later work showed that the amount of some other fertilizer, the depth of plowing, or the variety used in the experiment had been far from the best, unless there were strong reasons for believing that a change to optimal conditions in respect of these factors would not appreciably affect the needs of phosphate. Since agricultural experi-

Agricultural Research

ments take many months to perform and their evidence is scarcely trustworthy unless averaged over several seasons, the chain of experimentation required for adjusting factors to their optimal states one by one would continue for many years.

The alternative of planning experiments for the simultaneous study of several factors, each level or state of one being applied in combination with various levels of the others, enables far more rapid progress to be made. Plan 6.1 shows the design of a typical factorial experiment from an agricultural research station; it is presented without comment, but the reader should try to understand its structure when he has mastered later sections of this chapter. To many, the chapter will prove difficult, but the ideas are so important to a full appreciation of experimental design that it should be read carefully, and the reader should exercise himself with pencil and paper in constructing the designs mentioned.

6.2. FACTORIAL DESIGN IN OTHER SCIENCES

In certain other branches of science, the fact that experiments are often completed much more rapidly than in agriculture may modify the argument that complex factorial designs are essential if any progress is to be made in a short time, but the need to understand the dependence of one factor upon others remains. The three main reasons for including levels of several factors in one experiment are: (i) to obtain information on the average effects of all the factors economically, from a single experiment of moderate size; (ii) to broaden the basis of inferences on one factor by testing it under varied conditions of others; and (iii) to assess the manner in which the effects of factors interact with one another. These are not entirely independent, but the emphasis varies with the subject of experimentation.

Factorial Experiments

Fisher (1951, § 37) has stated the case for factorial experiments with great clarity. He says:

We are usually ignorant which, out of innumerable possible factors, may prove ultimately to be the most important, though we may have strong presuppositions that some few of them are particularly worthy of study. We have usually no knowledge that any one factor will exert its

PLAN 6.1
FIELD PLAN OF A SUGAR-BEET EXPERIMENT

The experiment consisted of 4 blocks of 16 plots. The symbols represent:

d: dung, at 10 tons per acre,
p: superphosphate, at 0.5 cwt. P_2O_5 per acre,
k: muriate of potash, at 1.0 cwt. K_2O per acre,
s: agricultural salt, at 5 cwt. per acre.

The dressings of p, k, s were applied together at one of four times, symbolized by

1: broadcast in November and plowed under in January,
2: broadcast in February,
3: broadcast in March,
4: broadcast at sowing, in May.

The plan shows the relative positions of plots, but is not to scale. Roman numerals denote the blocks, bounded by full lines; broken lines separate the plots.

	ds1	ps2	pk3	ps4	p2	d	d	pks1	
	pk1	dpks1	ds3	Nil	p4	dpk1	dks2	pks3	
I	ks2	dk4	dk2	dp4	s1	dps2	dks4	dpk3	II
	dpks3	dp2	ks4	Nil	s3	dps4	k2	k4	
	ks3	ds2	dk1	Nil	s4	dks3	dks1	pks2	
	dpks2	pk4	ps1	dp3	dps1	p1	p3	dpk4	
III	dk3	Nil	ds4	pk2	d	dpk2	dps3	s2	IV
	ps3	dp1	dpks4	ks1	d	pks4	k3	k1	

Certain interactions of the treatment factors are confounded between blocks (see § 6.10). (Rothamsted Experimental Station, 1938, p. 147.)

effects independently of all others that can be varied, or that its effects are particularly simply related to variations in these other factors. On the contrary, when factors are chosen for investigation, it is not because we anticipate that the laws of nature can be expressed with any particular simplicity in terms of these variables, but because they are variables which can be controlled or measured with comparative ease. ... The modifications possible to any complicated apparatus, machine or industrial process must always be considered as potentially interacting with one another, and must be judged by the probable effects of such interactions. If they have to be tested one at a time this is not because to do so is an ideal scientific procedure, but because to test them simultaneously would sometimes be too troublesome, or too costly. In many instances ... the belief that this is so has little foundation. Indeed, in a wide class of cases an experimental investigation, at the same time as it is made more comprehensive, may also be made more efficient if by more efficient we mean that more knowledge and a higher degree of precision are obtainable by the same number of observations.

6.3. Example of a Factorial Experiment

A factorial experiment is usually (but not necessarily: §§ 6.9, 7.3) one in which several states of two or more factors are tested in all possible combinations. As a prelude to an account of these designs, an experiment in which the principle was used with great success will be described; it illustrates how a well-designed experiment, even when of highly complex factorial design, can manifest its main conclusions without any great amount of calculation.

Kalmus (1943) studied the constitution of Pearl's synthetic medium for a yeast culture of *Drosophila*. In addition to agar, cane sugar, and tartaric acid, the medium advocated by Pearl contained the following ingredients:

		Per Cent
(N)	Ammonium sulphate, $(NH_4)_2SO_4$........	0.2
(M)	Epsom salt, $MgSO_4 \cdot 7H_2O$.............	0.05
(C)	Calcium chloride, $CaCl_2$................	0.025
(K)	Rochelle salt, $KNaC_4H_4O_6 \cdot 4H_2O$........	0.8
(P)	Primary potassium phosphate, KH_2PO_4..	0.1

Factorial Experiments

The obvious way of investigating the efficacy of this medium would be to compare cultures bred on it with cultures from media containing these or other salts in various proportions. One medium might be shown to be markedly superior, but, unless the alternatives had been carefully chosen, the cause of its superiority would probably remain in doubt. Kalmus restricted his attention to the five salts and sought to examine whether all were necessary or whether some might not even be harmful. He prepared 16 media, alike in all respects except the salts, having each of the possible combinations of absence of one of N, M, C, K or presence at Pearl's percentage, all being without potassium phosphate; thus the combination MK contained only Epsom and Rochelle salts. Two other series of 16 media had the same combinations of N, M, C, K with P at 0.05 and at 0.15 per cent, thus bracketing Pearl's recommendation. He made up four vials with each of the 48 media, placed two male and three female *D. melanogaster* in each for a week, and counted the hatch of flies.

The mean numbers of flies per vial, averaged without regard to the combinations of M, C, K, P, were:

    96 vials without N........................ 0.5
    96 vials with 0.2 per cent N................ 15.9

Similar averaging without regard to N, M, C, K showed:

    64 vials without P........................ 0.02
    64 vials with 0.05 per cent P............... 9.9
    64 vials with 0.15 per cent P............... 14.7

Ammonium sulphate and potassium phosphate are clearly essential if any reasonable number of flies is to be hatched, and analysis is hereafter restricted to the 64 vials that had both of these ingredients. Further averaging of eight groups of 8 vials gave the results in the accompanying tabulation. Comparison of the two entries in each column shows that, in general, Rochelle salt was seriously detrimental, the only exception being where yields were in any case low. Epsom salt was consistently beneficial, and calcium chloride showed no very clear effects. Moreover, averaging over all combinations

	No M, No C	0.05 Per Cent M, No C	No M, 0.025 Per Cent C	0.05 Per Cent M, 0.025 Per Cent C
No K.............	23.9	42.8	12.6	44.1
0.8 per cent K.....	7.5	24.2	14.6	20.9

of C and K, the means of sets of 16 vials suggest that the advantage from the larger amount of potassium phosphate appears only when Epsom salt is supplied (see table). These

	No M	0.05 Per Cent M
0.05 per cent P........	14.2	23.8
0.15 per cent P........	15.1	42.2

later conclusions are less clearly established from inspection of means than were those relating to the necessity for ammonium sulphate and potassium phosphate: an analysis of variance is needed in order to provide tests of significance. Nevertheless, inspection strongly indicates that Rochelle salt should be and calcium chloride might be omitted. The analysis of variance is required only to give objectivity to inferences that good experimental planning has made apparent with the aid of nothing more than totaling and averaging.

Of course, the study is not completed by this one experiment. Kalmus pointed out that a further experiment was needed in order to examine the effects of different nonzero amounts of N, M, and P. He later made such an experiment,

Factorial Experiments

using the 27 combinations of 0.3, 0.4, and 0.5 per cent N; 0.08, 0.16, and 0.24 per cent M; and 0.2, 0.3, and 0.4 per cent P; these levels were perhaps excessively high relative to those previously tested, and no significant increases in the hatch of flies were obtained.

6.4. SPECIFICATION OF FACTORIAL DESIGNS

The design of Kalmus's experiment is described as a $2 \times 2 \times 2 \times 2 \times 3$, or $2^4 \times 3$ factorial in 4 replicates: it contains four factors (N, M, C, K) each at two levels (zero and another) and one factor (P) at three levels, so that there are, in all,

$$2^4 \times 3 = 48$$

possible combinations of levels to be tested. The experiment in Plan 6.1 was a $2^4 \times 4$ factorial (though with one or two complicating features), the four manurial factors being tested each at two levels and the factor relating to time of application of the inorganic manures at four levels. The term *level* is customary general terminology even when the comparison is between qualitatively different states of a factor. For example, if Kalmus had included a comparison between four different types of vial, this would have been an additional factor at four levels.

In theory, a factorial design can involve any number of factors at any number of levels, such as a $2 \times 3^3 \times 5 \times 8 \times 10^2$ involving 216,000 treatment combinations! In practice, limitations of time and resources exclude the more extravagant possibilities, and skill is needed in order to find a design conforming to an over-all restriction on size as well as to other constraints imposed by the subject matter of the experiment. For reasons that will appear, the two most widely used classes of design are 2^n and 3^n, n factors each at two or three levels, values of n ranging from 2 up to perhaps 7 or 8.

Specification of Designs

The first class can be modified to include a factor at four levels by regarding these as the combinations of two quasi-factors at two levels. Similarly, a factor at eight levels can be regarded as three such quasi-factors, and one at nine levels as two quasi-factors within the 3^n system. Quasi-factors require caution in interpretation. Designs like 5^n are rarely used because the number of treatments is so large even for $n = 3$. *Mixed designs*, in which not all factors have the same number of levels, are also used, important ones being the various simple combinations of 2 and 3: 2×3, $2^2 \times 3$, 2×3^2, $2^2 \times 3^2$, and so on; these, however, can usually be less satisfactorily fitted to the requirements of an experiment, and their statistical analysis is more laborious (§ 6.11).

The epithet "factorial" relates only to the relationships among the treatments. When the whole set of treatments has been specified, any of the schemes of chapters iv and v may determine the allocation to plots, completely randomized (as in Kalmus's experiment) and randomized block designs being common. Because of the large numbers of treatments to be included in one experiment, special types of incomplete block design are exceedingly important (§§ 6.10–6.13).

6.5. A 2^3 Experiment

Potter and Gillham's investigation (1946) of the toxicity of a pyrethrins spray to *Tribolium castaneum* used a simple factorial design. In order to examine the effect of storage conditions, tests were made on insects that, before spraying, had been stored in cool or in hot conditions; after spraying, each level of this factor was subdivided for further storage in cool or hot conditions until the assessments of mortality were made. These four combinations were repeated with the addition of terpineol to the spray. With each of the eight (2^3) treatments, several concentrations of spray were tried, and

Factorial Experiments

the median lethal concentration (§ 5.3) was estimated. Table 6.1 shows that, in either period of storage, cool conditions made the spray more toxic than did hot conditions; the effect was particularly great in the postspray period. The experiment also brought out information that no nonfactorial design could have given, namely, that, although the addition of terpineol had little average effect (potency relative to "no terpineol" slightly less than unity), the contrast between the potencies under cool and under hot storage was much more

TABLE 6.1

RELATIVE POTENCIES OF A PYRETHRINS SPRAY
(I.e., Ratio of Equally Effective Concentrations)

COMPARISON	MEAN	BEFORE SPRAYING		AFTER SPRAYING		TERPINEOL	
		Hot	Cool	Hot	Cool	Absent	Present
Cool vs. hot before spraying..........	1.39	......	......	1.43	1.35	1.19	1.63
Cool vs. hot after spraying..........	3.16	3.24	3.07	......	......	2.25	4.43
Terpineol vs. no terpineol............	0.94	0.80	1.09	0.67	1.31	......	......

marked when terpineol was added to the spray: without terpineol, cool storage after spraying gave the spray 2.2 times the toxicity that it had under hot storage, but with terpineol this factor became 4.4 (Finney, 1952a, § 51).

6.6. NOTATION

Each factor in an experiment is labeled with a Roman capital, either chosen to suggest the nature of the factor (D, P, K, S, T in Plan 6.1) or arbitrarily A, B, C, ... The levels are then symbolized by the corresponding lower-case letters with subscripts 0, 1, 2, ... ; for a quantitative factor, 0 would correspond to the lowest level (whether or not this were zero), and for a factor representing purely qualitative comparisons, the allocation of subscripts would be arbitrary. Thus $a_2b_0c_3d_3$ would represent a treatment combination in a four-factor experiment with factor A at level 2, B at level 0, and C and D both at level 3. For factors at 2 levels, level 1 can be symbolized more concisely by a letter without subscript (simply "a") and level 0 by absence of any symbol for that factor: acd would represent level 1 of A, C, and D, with level 0 of B. The combination of level 0 of every factor in a 2^n experiment is denoted by "(1)" or simply "1." The same practice can be usefully adopted, however many levels a factor has, but this is less usual.

6.7. Analysis of Variance

The statistical analysis of a factorial experiment follows the lines of §§ 4.8 and 4.11, but the sum of squares for treatments can be subdivided into components representing differences associated with particular factors or groups of factors. If a factor is tested at p levels, the degrees of freedom for treatments will include $(p - 1)$ for differences between the mean values of the observations at those levels; a sum of squares corresponding to these can be separated from the whole sum of squares for treatments and examined as representative of the *main effect* of the factor. If two factors have p and q levels, the degrees of freedom for treatments will include $(p - 1)(q - 1)$ relating to the manner in which the effect of one factor varies from one level to another of the second, and a corresponding sum of squares can again be isolated. This *two-factor* or *first-order interaction* is a symmetrical property of the factors; it can equally well be regarded as relating to the manner in which the effect of the second factor depends upon the level of the first. Similarly, if a third factor has r levels, one can find a sum of squares with

Factorial Experiments

$(p-1)(q-1)(r-1)$ d.f. for the three-factor or second-order interaction. In particular, in a 2^n design, every main effect and interaction has 1 d.f. This subdivision of the sum of squares for treatments is made possible by having equal numbers of plots of every treatment combination, in consequence of which the contrasts between plots corresponding to each main effect or interaction are orthogonal (§ 4.12) with those for every other main effect or interaction.

The main effects are symbolized by the letter for the factor, and the interactions by the appropriate sets of letters, written $A \times B \times D$, A.B.D, or simply ABD. Table 6.2 shows how to set out a complete analysis of variance for Kalmus's experiment (§ 6.3), on the assumption that the allocation of treatments to vials was completely randomized; if replicate sets of 48 vials had been assigned to different incubators, 3 d.f. for the four blocks would have been removed from the error component. The almost total failure of cultures without ammonium sulphate or potassium phosphate indicated that the analysis ought really to be restricted to 64 vials in a 2^4 design.

6.8. REPLICATION

The experimenter who learns to appreciate the advantages of factorial experiments will soon find his fertility of imagination in thinking of factors outstripping his powers of performing the experiments. An investigation in which simultaneous study of 6 factors seemed desirable would not be exceptional, but with each factor at two levels, it would involve 64 treatments, and with each at three levels, 729 treatments; replication of the first might be practicable, but few could seriously consider replicating a set of 729 treatments.

The way of avoiding this difficulty is to omit true replication! Interactions of four or more factors will usually be neg-

Replication

ligible, at least when the experimenter knows enough about his factors to be able to avoid including catastrophic combinations of levels. When a particular interaction is in reality zero—that is to say, the magnitude of the lower-order interaction between all but one of its factors is unaffected by the

TABLE 6.2

OUTLINE OF ANALYSIS OF VARIANCE FOR
EXPERIMENT OF § 6.3

Source of Variation	d.f.	Sum of Squares	Mean Square
N	1		
M	1		
C	1		
K	1		
P	2		
NM	1		
NC	1		
NK	1		
NP	2		
MC	1		
MK	1		
MP	2		
CK	1		
CP	2		
KP	2		
NMC	1		
NMK	1		
NMP	2		
NCK	1		
NCP	2		
NKP	2		
MCK	1		
MCP	2		
MKP	2		
CKP	2		
NMCK	1		
NMCP	2		
NMKP	2		
NCKP	2		
MCKP	2		
NMCKP	2		
Error	144		
Total	191		

Factorial Experiments

level of the remaining factor—its mean square in the analysis of variance has the same expectation as the error mean square. Hence a mean square obtained by pooling the sums of squares and the degrees of freedom for several high-order interactions should approximate to the error mean square and may be used as such. Any true interaction will tend to inflate this mean square a little, but the fact that main effects and interactions of low order are being examined in relation to higher-order interactions rather than to error alone is likely to be of small importance by comparison with the advantage of keeping an experiment on many factors within reasonable limits; indeed, this is sometimes an advantage.

A 2^6 experiment could well be performed in 64 plots only; of its 63 d.f., 15 correspond to 4-factor, 6 to 5-factor, and 1 to the 6-factor interaction, and these 22 d.f. might be used to give an estimated error mean square. If there were a priori reasons for believing that one or two of the 4-factor interactions were of special interest, these could be kept apart in the analysis, since 22 d.f. are more than enough for a satisfactory estimate of error. In practice, even 3-factor interactions are often used for error: for a 2^5 experiment in 32 plots, 16 d.f. from 3-factor and higher-order interactions may be used as error, again with the possibility of separating for special examination any interactions believed likely to be important. Two other very important possibilities (see § 6.10) are the 3^3 and 3^4 in single replication, using, respectively, 8 d.f. from 3-factor interactions and 16 d.f. from 4-factor interactions as error.

A single-replicate factorial experiment does not offend against the requirements of replication stated in § 4.4. The 2^5 in 32 plots, for example, has 16-fold replication of each level of each factor separately; not only is the variation among these used to give the estimate of variance, but the main

Replication

effect of the factor is measured as precisely as if no other factors were included and the experiment consisted solely of two sets of 16 identically treated plots. Similarly, the experiment has 8-fold replication of every combination of levels of any pair of factors.

6.9. Fractional Replication

When the number of factors is large, even an experiment employing only a fraction of the possible treatment combinations may give useful information on all main effects and important interactions. This can be illustrated by a 2^4 design, although *fractional replication* is not practically important for so few factors. Suppose that measurements on one plot of each of a particular eight combinations of factors were as follows:

Treatment	1	d	ab	abd	ac	acd	bc	bcd
Measurement	y_1	y_2	y_3	y_4	y_5	y_6	y_7	y_8

The treatments have been carefully chosen to preserve some balance over the factors. The main effect of A, the mean difference between plots with a and plots without, will apparently be estimated by

$$A = \tfrac{1}{4}(-y_1 - y_2 + y_3 + y_4 + y_5 + y_6 - y_7 - y_8).$$

So, for B,

$$B = \tfrac{1}{4}(-y_1 - y_2 + y_3 + y_4 - y_5 - y_6 + y_7 + y_8),$$

with similar expressions for C and D. Consider now the interaction between A and B. The effects of A in the absence and in the presence of b are obtained from two groups of 4 plots each as

$$\tfrac{1}{2}(-y_1 - y_2 + y_5 + y_6)$$

and

$$\tfrac{1}{2}(y_3 + y_4 - y_7 - y_8),$$

95

Factorial Experiments

respectively. By definition, the interaction is half the difference between these ($\frac{1}{2}$ in order to put the value in units of measurement per single plot):

$$AB = \tfrac{1}{4}(y_1 + y_2 + y_3 + y_4 - y_5 - y_6 - y_7 - y_8).$$

Except for a change of sign, this is obviously also the expression for the main effect of C; in symbols

$$AB = -C.$$

Similarly, every main effect and interaction has an *alias:*

$$AC = -B,$$
$$BC = -A,$$
$$ABD = -CD,$$
$$ACD = -BD,$$
$$BCD = -AD,$$

and

$$ABCD = -D.$$

No analysis of the eight observations can distinguish between what is due to a main effect of A and what is due to an interaction between B and C. If the experiment had consisted solely of the other eight combinations of the factors, the same relationships would have held except for a change of sign. They arise because, in the formation of the ABC interaction from the 16 possible treatment combinations, the two sets of eight would require negative and positive signs respectively. Either set constitutes a half-replicate of the design, which may be symbolized by

$$ABC = 1.$$

This symbolism indicates that no estimate of the ABC interaction itself can be formed; also, any main effect or interaction has as its alias the effect obtained by writing its algebraic product with **ABC** and then omitting any letter that is "squared": thus

$$B = ABC \cdot B \quad = AB^2C \quad = AC,$$
$$D = ABC \cdot D \quad = ABCD,$$

and

$$ABD = ABC \cdot ABD = A^2B^2CD = CD,$$

with other relations as before (signs can be neglected).

This rule for fractional replication of 2^n designs applies generally. The positive terms in the **ABCD** interaction are

1, ab, ac, ad, bc, bd, cd, abcd;

and choice of these eight as a half-replicate would be symbolized by

$$ABCD = 1.$$

Of aliases then found,

$$A = BCD,$$
$$AB = CD,$$

are typical. Again, if only four treatments had been included in the experiment, say

1, ab, acd, bcd,

these are combinations that simultaneously receive a negative sign in **ABC**, a negative sign in **ABD**, and a positive sign in **CD**. For all other main effects and interactions, two of the four treatments are taken positively and two negatively. Symbolically,

$$ABC = ABD = CD = 1,$$

where, according to the *generalized product rule* given above, the product of any two is the third:

$$ABD \cdot CD = ABCD^2 = ABC.$$

Also, every effect now has three aliases; the reader should

97

Factorial Experiments

verify that both direct construction from the four treatments and application of the rule lead to

$$A = BC = BD \quad = ACD,$$
$$B = AC = AD \quad = BCD,$$
$$C = AB = ABCD = D.$$

The half- and quarter-replicate designs so far discussed are of no practical use, since they do not allow main effects and 2-factor interactions to be kept distinct. However, bigger experiments can be so arranged that no main effect or 2-factor interaction has an alias of lower order than 3- or 4-factor interactions, and any large effect found can then be correctly ascribed with near-certainty. For example, a 2^7 experiment might be performed in a half-replicate of 64 plots, by taking

$$ABCDEFG = 1;$$

any main effect has a 6-factor interaction and any 2-factor interaction a 5-factor interaction as its alias, and there would be little uncertainty in interpreting any effects that appeared in the analysis. The 3-factor interactions, 35 d.f., whose aliases are all 4-factor interactions, would be used for the estimation of error, except that, if the main effects and 2-factor interactions concerned in a particular 3-factor interaction were large, it could be kept apart from the error sum of squares and tested. Even a quarter-replicate of 2^8 can be accommodated on 64 plots, by taking

$$ABCDE = ABFGH = CDEFGH = 1.$$

A set of treatment combinations for a particular fractional replicate of 2^n is easily found. It consists of treatment "1" (i.e., the zero level of all factors) and every other combination having an even number of letters in common with each of the interactions defining the fraction (zero, of course, is an even number). Moreover, the generalized product rule helps the search for combinations: the product of any two symbols satisfying the condition, after omission of any letter that is squared, is also a member of the set. For example, for the quarter-replicate of 2^8 specified above, each of ab, cd, ce, fg, fh, and acf contains either 2 or 0 letters from ABCDE, ABFGH, and CDEFGH, and they and all products (such as abcd, c^2de = de, a^2bcf = bcf) that can be formed from any number of them, together with 1, give a set of 64 combinations. When one set is known, another can be generated by multiplication of each of its members by any one treatment combination not included in it; for half-replicate designs, the second half consists merely of the remaining combinations.

Fractional replicates of other designs are also important. A one-third replicate of 3^5 in 81 plots can be arranged so that main effects and 2-factor interactions have 4-factor and 3-factor interactions, respectively, as their aliases of lowest order. Provided that these higher-order interactions can reasonably be expected to be substantially smaller than main effects and 2-factor interactions, this design is useful for investigating the interrelationships of 5 factors within an experiment of reasonable size (see § 6.10). One-ninth replicates of 3^n are useful for larger values of n.

Fractional replication of mixed factorial schemes is not very satisfactory, except in so far as the fraction can be arranged to relate to factors at one level only. For example, a half-replicate of $2^6 \times 3$ in 96 plots might be constructed as 32 combinations for one-half of 2^6 combined with all levels of the other factor.

6.10. Confounding

The arguments advanced in § 4.7 for arranging treatments in blocks remain valid when the treatments have a factorial

Factorial Experiments

structure. If the total number of treatment combinations is small, factorial designs can be arranged as randomized blocks or Latin squares, but 12 or 16 combinations in randomized blocks and 8 or 9 in a Latin square are often about the largest numbers that can be satisfactorily accommodated. For larger numbers of combinations, the incomplete block designs of chapter v can be used. The factorial structure, however, gives opportunity for constructing incomplete blocks on an alternative principle, deliberately sacrificing precision on certain interactions in order that more important effects may be measured more precisely.

The simplest of examples is provided by a design for a 2^2 experiment in blocks of 2. If each replicate is divided into two blocks:

(i)	(ii)
1	a
ab	b

the difference "second plot minus first plot" from all blocks of type (i) added to the difference "first plot minus second plot" from all blocks of type (ii) leads to an estimate of the main effect of A (for it is balanced in respect of levels of B). Subtraction of the second difference from the first, symbolically

$$(ab - 1) - (a - b),$$

leads similarly to an estimate of the main effect of B.[1] Now these comparisons between plots involve one plot positively and one negatively from every block and are therefore orthogonal with all block differences. On the other hand, the AB interaction would have to be calculated from the total of blocks of type (i) minus the total of blocks of type (ii): the comparison of treatments required is identical with a differ-

---

1. These quantities must be divided by the total number of blocks, in order to give effects in units of one plot.

ence between two sets of blocks. In the terminology of § 6.9, AB may be said to have a block difference as an alias, but, where blocks are involved, it is more usual to say that the interaction AB is *confounded* with blocks. A symbol q can be used to represent the fact of a treatment going into the second type of block, the omission of q indicating the first type of block: the design then consists of repetitions of the treatments and block allocations specified by

$$1, \quad aq, \quad bq, \quad ab.$$

This has the form of the half-replicate specified by

$$ABQ = 1$$

for the combinations of levels of A, B, and a quasi-factor Q. As explained in § 6.9, this equation leads to

$$AB = Q,$$

the symbolic statement that AB is confounded with blocks.

The experiment just discussed is of restricted practical value, because it entails the sacrifice of information on the interaction AB, and this can rarely be tolerated. It is not entirely useless: if the treatments related to the manner of making virus inoculations and the plots of a block were two halves of one leaf, an average difference between leaves would estimate the interaction. For example, two leaves might be used on each of 12 plants, one leaf of each plant being chosen at random[2] as a block of type (i) and the other becoming a block of type (ii). Random halves of each leaf would then be assigned to one of the two treatments of the

2. If the upper leaf were always assigned to blocks of type (i), Q would represent a comparison of upper with lower and not merely a random comparison between leaves of the same plant; the alias statement $AB = Q$ could no longer justify the interpretation of any consistent block difference as in reality a result of interaction

Factorial Experiments

block. The main effects would be estimated with the precision of intraleaf variation, the interaction possibly much less precisely from interleaf variation (cf. §§ 6.12, 8.4).

When many factors are involved, the potentialities of confounding are greatly increased. For example, a 2^5 design can be arranged in blocks of 16 by confounding the 5-factor interaction, usually a small sacrifice, since this is rarely of much interest. By confounding two 3-factor interactions simultaneously, such as ABC and ADE, the blocks are reduced to 8 plots. Plan 6.2 shows a single replicate of this scheme, which

PLAN 6.2

DESIGN FOR 2^5 EXPERIMENT IN BLOCKS OF 8, CONFOUNDING ABC, ADE, BCDE

I	II	III	IV
1	b	d	bd
bc	c	bcd	cd
de	bde	e	be
bcde	cde	bce	ce
abd	ad	ab	a
acd	abcd	ac	abc
abe	ae	abde	ade
ace	abce	acde	abcde

can be repeated as often as desired with fresh randomization of order within each block. A consequence of confounding two interactions is that the generalized product of their symbols (§ 6.9) is also confounded: the product of each pair of confounded interactions is the third. The reader may verify that the difference between blocks of Types I and II and those of Types III and IV corresponds to ADE; that I and III versus II and IV corresponds to ABC; and that I and IV versus II and III corresponds to BCDE. Moreover, the first block consists of all the treatment combinations having an even number of letters from each of the sets a, b, c; a, d, e; and b, c, d, e; and the other blocks are generated from it by generalized multiplication with b, d, and bd, respectively. These properties, closely connected with similar properties of fractional replication, are important in the construction of confounded designs (Fisher, 1942; Finney, 1947).

Fisher (1942) has proved that even a 2^7 design can be arranged in blocks of 8 plots without confounding any main effects or 2-factor interactions: 15 interactions of second and higher order are then confounded, these having the property that the generalized product of any two is a third. With

PLAN 6.3

ARRANGEMENT OF A 3^3 DESIGN IN
3 BLOCKS OF 9

Combinations of First Two Factors		Blocks		
a	b	I c	II c	III c
0	0	0	1	2
1	0	1	2	0
2	0	2	0	1
0	1	1	2	0
1	1	2	0	1
2	1	0	1	2
0	2	2	0	1
1	2	0	1	2
2	2	1	2	0

blocks of 16, up to 15 factors can satisfy the same restriction on confounding.

Provided that enough high-order interactions remain unconfounded and are suitable for the estimation of error, single replicates of confounded factorial designs can be used. A very valuable scheme is that for a 3^3 design in 3 blocks of 9, confounding 2 d.f. out of the 8 d.f. for ABC; Plan 6.3 shows one of the four possible arrangements. The analysis of variance is first made in the form of Table 6.3, and the error mean square is then based upon the 6 unconfounded degrees of freedom for ABC with the addition of any from the 2-factor interactions that seem of least interest. This and a similar confounding for 3^4 in 9 blocks of 9 are of immense practical value in the many problems for which inclusion of more than two levels of a factor is essential.

All confounded factorial designs can be regarded as fractional replicates of schemes in which one or more quasi-fac-

TABLE 6.3

OUTLINE OF ANALYSIS OF VARIANCE OF
EXPERIMENT IN PLAN 6.3

Source of Variation	d.f.	Sum of Squares	Mean Square
Blocks	2		
A	2		
B	2		
C	2		
AB	4		
AC	4		
BC	4		
ABC (unconfounded)	6		
Total	26		

tors represent the comparisons between blocks. Often, however, the ideas of pure fractional replication and of confounding can profitably be combined, giving a design that provides information on all the more important effects without testing all possible combinations, yet that can be executed in blocks of moderate size. Thus a half-replicate of 2^7 can be arranged in 8 blocks of 8 plots (Plan 6.4). It is defined by

$$ABCDEFG = 1,$$

and the confounded interactions are ABD, ACE, CDG, AFG,

BCF, BEG, DEF, and their aliases.[3] A half-replicate of 2^6 can be arranged in 2 blocks of 16, but not in 4 blocks of 8 unless a 2-factor interaction is confounded. For experiments in which factors are tested at three different levels, fractional replication is even more important because of the large number of treatment combinations arising from only a few factors. Fortunately, satisfactory confounding schemes can be constructed for fewer factors than with 2^n. One-third of a

PLAN 6.4*

DESIGN FOR CONFOUNDING HALF-REPLICATE OF 2^7 IN BLOCKS OF 8

I	II	III	IV	V	VI	VII	VIII
1	ab	ac	ad	ae	af		
abcg	cg	bg					
abef	ef	bcef	\multicolumn{3}{l}{The reader should complete Blocks IV–VI himself by generalized multiplication of Block I by ad, ae, af, and should then find suitable multipliers to give Blocks VII and VIII}		
cefg	abcefg	aefg					
acdf	bcdf	df					
bdfg	adfg	abcdfg					
bcde	acde	abde					
adeg	bdeg	cdeg					

* NOTE: (1) Every treatment combination contains an even number of letters. (2) Every treatment combination in Block I contains an even number of letters from every confounded interaction. (3) The generalized product of any two elements in Block I is also in Block I. (4) Block II would be formed in a different order if any other of its treatments were written first and used in generalized multiplication of Block I: bcdf·abcg = adfg. So for other blocks.

replicate of 3^5 can be arranged in 9 blocks of 9 in such a way that all main effects and 2-factor interactions have higher-order interactions as aliases, and the only serious loss is that 2 d.f. from one 2-factor interaction must be confounded; Chinloy *et al.* (1953) illustrated the use of a less satisfactory variant in an experiment on the manuring of sugar cane. A more ambitious design used by Tischer and Kempthorne (1951) was a 3^7 in one-ninth replication, arranged in 9 blocks

3. Note that these triads of letters form a balanced incomplete block scheme for the seven letters (cf. Plan 5.2), an example of how apparently entirely different types of design can be linked.

Factorial Experiments

of 27; this great simplification in the problem of examining a potential total of 2,187 treatment combinations was entirely justified by the appearance of very few interactions.

6.11. Partial Confounding

Mixed designs (§ 6.4) cannot be confounded as easily as can the 2^n and 3^n types. Sometimes the confounding can be restricted to one set of factors, all with the same number of levels. For example, a 3×2^3 experiment might be put into pairs of blocks of 12 plots so as to confound the 3-factor in-

PLAN 6.5

Design for Confounding 3×2^2 in Blocks of 6

Combinations of First Two Factors		Blocks					
		I	II	III	IV	V	VI
a	b	c	c	c	c	c	c
0	0	0	1	0	1	0	1
0	1	1	0	1	0	1	0
1	0	0	1	1	0	1	0
1	1	1	0	0	1	0	1
2	0	1	0	0	1	1	0
2	1	0	1	1	0	0	1

teraction from the 2^3: in any one block, the same four combinations of these 3 factors would be associated with each level of the first factor. Alternatively, *partial confounding* can be adopted. A 3×2^2 experiment can be put into blocks of 6, by using three pairs of blocks, each of which forms a replicate (Plan 6.5), in such a way that the BC and ABC interactions are neither orthogonal with blocks nor identical with block differences. Six (or a multiple of 6) blocks are needed in order to balance the pattern of the confounding; provided that pairs of blocks are used, this restriction can be dropped at the price of extra complexity in an already laborious statistical analysis. The principle easily extends to 3×2^3 in blocks of

12. Fractional replication of this type of design appears to have little practical importance.

When a factorial experiment is to be confounded in order to keep the block size small but is to be replicated more than once, different interactions can be confounded in different replicates. In the virus experiment of § 6.10, if more information on AB were wanted, A, B, and AB might be confounded in equal numbers of replicates.[4] A 2^4 design in 8 blocks of 8 might have ABCD confounded in the first pair of blocks, ABC, ABD, and ACD in the others, so enabling these effects to be estimated, albeit with lower precision, from the blocks in which they are unconfounded. A 3^3 design in blocks of 9 might be arranged in 12 blocks, confounding a different pair from the 8 d.f. for ABC in each of four replicates. This type of design, also referred to as partial confounding, has no merits unless the experimenter is seriously interested in the interactions concerned; otherwise, replication of a completely confounded scheme is equally good and easier in analysis (§ 9.2).

6.12. SPLIT-PLOT DESIGNS

Occasionally some factors in an experiment can be applied differentially to smaller units than can others. Dietary comparisons must be made on whole animals, whereas drugs can sometimes be compared by injection at different sites on one animal. Factors relating to the sources of seeds must affect whole plants, but virus inoculations can be compared on leaves or half-leaves of a plant. The comparison of soil-cultivation techniques that employ unwieldy implements may demand large plots, but tests of fertilizers or other agronomic factors may be made simultaneously on subdivisions of these areas. An experiment in which some treatments are applied to large units, or *main plots*, each of which is divided into two

4. This arrangement is also a balanced incomplete block design!

Factorial Experiments

or more *subplots* for other treatments, is said to have a *split-plot* design. The principle is simply that certain main effects and their interactions with one another are confounded (main plots corresponding to blocks and subplots to plots). The emphasis is shifted, however, since an ordinary confounding design is usually planned with the intention of obtaining no information on certain interactions, whereas a split-plot design must have sufficient replication of main plots to give adequate precision on main-plot factors.

Splitting of plots can be used to introduce an extra factor into an experiment that is in progress. In agricultural or other research that continues over a long period, this is useful for allowing new ideas to be incorporated, although it increases the number of plots. The possibility of introducing the new factor by applying different levels to whole plots, in accordance with an extended confounding scheme, should always be examined as an alternative that demands no increase in plots. For example, conversion of a single replicate of 3^4 into one-third of 3^5 might often be preferable to the increase from 81 to 243 plots that splitting would necessitate. In short-term experiments, initial good planning will usually eliminate any need for modifications later.

Split-plot experiments will usually assess the effects of subplot factors and their interactions with main-plot factors more precisely than the effects of main-plot factors alone. Split-plot designs are therefore sometimes adopted in order to obtain higher precision on comparisons of greater importance: however, when no other considerations also favor split plots, a design confounding high-order interactions rather than main effects is often better still. In some fields of research, split plots are too commonly used without thought of whether the same object could have been better achieved in other ways. To arrange a factorial experiment with factor

A on main plots, these split into subplots for B, and these further split into sub-subplots for C, is easy but rarely gives the best design.

6.13. DOUBLE CONFOUNDING

By confounding one set of interactions with rows and another with columns, or with two orthogonal systems of blocks analogous to rows and columns, the advantages of Latin square designs can be brought into factorial experimentation. The *plaid squares*, obtained when certain main effects are confounded with rows or columns or both, are a form of such *double confounding:* these have all plots of one row or column at the same level of a factor, so that they have some of the operational advantages of split-plot designs. Double confounding requires care if invalid and unsatisfactory designs are to be avoided.

6.14. DESIGNS FOR ZERO INTERACTIONS

Occasionally the main effects of factors can be assumed to be perfectly additive (i.e., all interactions zero). For example, the true weight of two articles in combination must be the sum of their separate weights; observations on weights will be subject to random errors and perhaps to systematic deviations from truth, but, over a small range of weight on a good balance, the latter ought to be negligible. Suppose that the weights of three objects are to be determined. The obvious course is to make four weighings, one with an empty pan to give a zero correction and one with each article in turn. If σ is the standard deviation of random errors for a single weighing, the standard error of the weight estimated for each article is $\sigma\sqrt{2}$.

Yates (1935) suggested an alternative procedure. If the first weighing is made with all three articles together (w_1) and the others with the articles a, b, c separately (w_2, w_3, w_4), the

Factorial Experiments

reader will easily see that, whatever the zero correction, the weights of the articles are estimated by

$$A = \tfrac{1}{2}(w_1 + w_2 - w_3 - w_4),$$
$$B = \tfrac{1}{2}(w_1 - w_2 + w_3 - w_4),$$
$$C = \tfrac{1}{2}(w_1 - w_2 - w_3 + w_4).$$

The standard error of each estimate is now only σ. A further improvement will be effected if, for the second, third, and fourth weighings, the other two articles can be put on the opposite pan of the balance, so that w_2 now measures the difference in weight between (a + zero correction) and (b + c). The same expressions give the estimated weights of the articles, except that the factor $\tfrac{1}{2}$ is replaced by $\tfrac{1}{4}$, and the standard error is now $\sigma/2$. The weights are thus determined much more precisely by no extra labor except that of organization. With larger numbers of articles, more substantial gains can be made. The reader may verify that, whereas 11 articles would have their weights determined with standard error $\sigma\sqrt{2}$ if each were weighed separately, the scheme shown in Plan 6.6 leads to estimates with standard error $\sigma/\sqrt{3}$ if articles are placed only in one pan or $\sigma/2\sqrt{3}$ if the articles not in one pan are always put in the other. This is one of many plans developed by Plackett and Burman (1946); its close relation to the balanced incomplete block design for $v = b = 11$, $k = r = 6$ mentioned in § 5.3 should be noted (the + signs in columns 2–12 give this design).

These designs are particular types of fractional replication of 2^n available when interactions can be completely ignored. Similar schemes can be constructed for 3^n. Their use seems likely to be greater in industrial research than in biology.

6.15. Information from Factorial Experiments

In § 6.2, three reasons for using factorial designs were stated. Although a factorial experiment may require more plots than would an experiment on any one of its factors alone, it will often be smaller than the totality of these separate "simple" experiments. Plan 6.2 illustrates this point. An experiment on any one of the five factors alone could be put on randomized blocks of 2 plots, and the standard deviation

PLAN 6.6

DESIGN FOR WEIGHING 11 ARTICLES IN 12 OPERATIONS

ARTICLE No.	OPERATION No.*												
	1	2	3	4	5	6	7	8	9	10	11	12
1....	+	−	−	+	−	−	−	+	+	+	−	+
2....	+	−	+	−	−	−	+	+	+	−	+	−
3....	+	+	−	−	−	+	+	+	−	+	−	−
4....	+	−	−	−	+	+	+	−	+	−	−	+
5....	+	−	−	+	+	+	−	+	−	−	+	−
6....	+	−	+	+	+	−	+	−	−	+	−	−
7....	+	+	+	+	−	+	−	−	+	−	−	−
8....	+	+	+	−	+	−	−	+	−	−	−	+
9....	+	+	−	+	−	−	+	−	−	−	+	+
10....	+	−	+	−	−	+	−	−	−	+	+	+
11....	+	+	−	−	+	−	−	−	+	+	+	−

* + = article put on left-hand pan; − = article omitted or put on right-hand pan.

per plot would no doubt be smaller than for blocks of 8: nevertheless, 6–8 replications would be the minimum that could be contemplated for an experiment to give the same precision for the effect of A as does the 16-fold replication in Plan 6.2. Repetition of this for each factor would use 60–80 plots, instead of 32.[5] Moreover, for each of these experiments, a choice would have to be made of the levels at which the other

5. Even if a set of 6 treatments were arranged in randomized blocks of 6, the treatments being chosen to test each factor separately (in symbols, 1, a, b, c, d, e, or perhaps 1, a, ab, abc, abcd, abcde, for the 6 treatments), 6-fold replication would require 36 plots and would give no information on interactions.

111

Factorial Experiments

four factors should be held; consequently, if the experiment on factor B were performed with E at the higher of its two levels, and the experiment on E were then to show the superiority of the lower level, the value of the experiment on B would be much reduced.

A factorial design, in fact, is an excellent insurance policy. If for Plan 6.2 the effect of each factor is independent of the levels of other factors, the five factors have their average effects measured in the experiment, each with the precision of 16-fold replication (in blocks of 8). If the effect of one factor is modified by the levels of others, the experiment gives an opportunity of detecting this interaction and of estimating its magnitude. An experimenter who is certain that he is interested in the effect of B only at the upper level of E may reasonably decline to include the lower level of E in his design; if he is unable to dismiss the possibility that the ideal treatment may involve any of the four combinations of levels of B and E, it is hard to see how he can reach a satisfactory decision otherwise than by factorial design (cf. § 7.3).

Essentially the same arguments hold for factors at three or more levels. When circumstances justify the risk of some confusion on high-order interactions, fractional replication enables an even larger number of factors to be included in one experiment, and the advantage to the economy of experimentation can be substantially greater than with single replicates.

CHAPTER VII

Sequential Experiments

7.1. Sequential Nature of Research

In the study of quantitative properties of living matter, attainment of a final and complete conclusion at the end of an experiment is exceptional. This is evident in applied science, where the empirical character of many results of practical importance is reason for neither obtaining nor demanding absolute accuracy; the "best" combination of fertilizers for growing wheat or the "best" hospital regime for the cure of tuberculosis is an ideal that can be realized only for particular concomitant circumstances, and even then experimental search for the best can do no more than give an approximation to the ideal. In pure science, some quantitative properties lend themselves to exact determination (for example, the number of chromosomes characteristic of a species), but again exactness is commonly unattainable; improved and enlarged experiments may estimate with increasing accuracy the relationship between temperature and the fertility of an insect, the relative potencies of different natural sources of a drug, or the frequency of chromosomal recombinations between two loci in a plant species, but will never lead to exact knowledge of these quantities.

Hence much biological research is necessarily *sequential*, in the sense that the results of one experiment are likely to be used as a basis for planning future experiments on the same topic (in addition to any immediate use that is made of them in advancing theory or improving practice). Designs that have been recently developed carry this idea further by permitting information accumulated during the progress of one experiment to be used in modifying the subsequent conduct of that experiment. Although these are not yet used extensively in biological research, the experimenter ought to be aware of some of their potentialities. Statistical theory in this field continues to develop rapidly, and only a brief review of four distinct types of experiment can be given here.

7.2. Factorial Experiments

The ambitious and imaginative experimenter who has learned to appreciate factorial designs may often discover that, despite the power of confounding and fractional replication, a single experiment cannot include all the factors and levels that interest him. If he has a limited total number of "plots" or other units at his disposal, but not all of these need be used simultaneously, and if results of the treatments applied to some can be available before others are treated, he may consider testing one set of factors in the early stages and then modifying the choice and the levels of factors for the later part of the experiment. Davies and Hay (1950) have suggested that a first stage might consist of a small fraction of a replicate of a factorial scheme for factors believed unlikely to have interactions. Even 10 factors each at two levels might be put on 16 plots so as to leave some degrees of freedom for estimating error; if interactions are feared, fewer factors can be included, but as many as 8 factors can still be arranged so that main effects have 3-factor and higher-order aliases, while 2-factor interactions are aliases of one another in sets of 4. The results of this fraction may then suggest that some factors be discarded as uninteresting, that levels of others be modified to more interesting values, and perhaps

Factorial Experiments

that new factors be brought in; alternatively, if no change seems desirable, another fraction of the whole replicate can be added.

Greater flexibility of design is thus retained, as the experimenter does not need to restrict himself to a choice of treatments made at the beginning of the experiment. Nevertheless, he runs the risk of missing important interactions or discarding interesting factors because their effects in the first stage were obscured by interactions. The method is perhaps more suited to technological research than to pure science, since it allows emphasis to be placed on finding the factors of greatest practical importance rather than on studying an arbitrarily selected set of factors. Floyd (1949) has described a simple application in connection with penicillin production and use.

7.3. EXPERIMENTAL SEARCH FOR OPTIMAL CONDITIONS

Important ideas have recently been put forward (Box and Wilson, 1951) for experiments whose object is to discover the combination of conditions that maximizes a yield or other assessment of performance. These have arisen in relation to industrial experimentation, where the combination of physical conditions (temperature, pressure, amounts and concentrations of different ingredients, time allowed for reactions, etc.) that maximizes the yield or the net return from some product is required. The generally lesser stability of conditions producing maxima in biological phenomena (because of extraneous uncontrolled factors) makes doubtful whether the methods will find much application in biology. Nevertheless, they are so interesting that a brief account ought to be given.

The principle is simple. The reader should have no difficulty in visualizing the process when only two factors are involved, even though he may have no idea of the mathematical technique required at each stage. The relationship between the average yield (or any other quantity under study) and the levels of two different factors can be represented by a relief map in which rectangular co-ordinates in a horizontal plane represent levels of the two factors and height represents the yield. The aim of the experiment is to estimate the levels of the factors that correspond to the highest point. The procedure may be expressed nonmathematically as follows:

i) Guess the required combination of levels, and measure yields for it and for a few other combinations differing slightly from it.

ii) Estimate the direction on the map in which yield increases most steeply from the point first guessed.

iii) Take new levels of the two factors a fixed short distance in this direction.

iv) As a second stage of the experiment, make tests of this new combination and of a few others differing slightly from it.

v) Repeat steps ii–iv until a combination of levels is reached at which the surface is found to rise to only a negligible extent in every direction.

On an average, the yield must increase as this process continues, though four dangers are present:

a) Experimental errors that are large relative to the differences in yield used in estimating slopes will make progress slow, because the direction taken will often differ considerably from the steepest slope.

b) The optimal levels of the factors may change during the course of the experiment because of difficulties in keeping other conditions fixed.

c) Within the region explored, the map may contain more

Search for Optimal Conditions

than one mountain peak, and the mountain that is climbed may not be the highest of all.

d) The process may end if an almost horizontal plateau is reached, whether or not the mountains rise above this.

Both danger *a* and danger *b* are likely to be encountered in biology, and either makes the situation scarcely suitable for this technique; results that are more reliable, though less ambitious in aim, will be obtained from the classical type of factorial design (chap. vi). Theoretical knowledge of the effects of the factors or a preliminary survey over a wide range of levels may serve to eliminate *c*, and mathematical refinements help to overcome *d*.

The generalization of this method to the simultaneous study of several factors complicates the mathematics but leaves the principle unaltered. Box and Wilson have made recommendations on the number of different combinations to be tested at each stage and the arrangement of these, as well as on other questions relating to the optimal designs. They show that the improvement in the economy of the experiment may be considerable, because expenditure of effort on combinations of levels known to be far from the optimal is saved. This consideration does not affect the importance of classical factorial designs in research into the relationship of yield to levels of factors over a wide range, but it may be very valuable in technological problems where interest is practically restricted to the optimal.

7.4. SEQUENTIAL RULES FOR TERMINATING EXPERIMENTS

Any experiment in which the conduct of one stage is determined by the results of earlier stages is properly styled sequential, but the growth of ideas on incorporating results into rules for conducting the experiment has been particularly important in circumstances where termination of the experiment rather than choice of treatment is sequentially determined. Once again the chief uses in the past have been industrial, but methods of this group will be illustrated here by reference to clinical experiments.

As emphasized in § 2.10, in the development of a new remedy for a disease a stage must be reached at which the new method is deemed safe for trial but each patient on whom it is tried is necessarily experimental. The obvious procedure for making a reliable comparison between a standard remedy, A, and a suggested improvement, B, would be: "Make a random selection of half the available patients for B, give A to the others, and after a suitable time examine the proportions cured." If the total number of patients wanted was not available at the start, pairs might be made up as patients were diagnosed, one of each pair being assigned to B and the other to A; in some circumstances, the pairs might be chosen alike in sex and might be further balanced in respect of age, severity of disease, or other characteristics. The pairing would eliminate any biases arising from secular trends in diagnosis or in the administration of treatments and the care of patients.

This use of a time sequence of pairs suggests a sequential design. If the results for any subject are obtainable fairly rapidly, any large difference in effectiveness of A and B is likely to betray itself from tests on only a few pairs: to continue until a preassigned number has been tested not only seems uneconomic experimentation but also offends against the ethical principle that a remedy shall not be used after it has been proved inferior. On the other hand, if the difference between A and B is small, a preassigned number of subjects may fail to point decisively to either as the better, and to stop the experiment at that total could be almost equivalent to wasting all the work already done. In practice, most clinical experimenters no doubt decide whether to continue or to end an experiment from study of the results already obtained, and what is wanted is an objective rule of conduct.

Bross (1952) discussed this problem in the light of statistical theory developed earlier for analogous situations. As results for pairs of patients accumulate, they can be classified into four groups: (i) neither cured; (ii) A cured, B not cured; (iii) A not cured, B cured; (iv) both cured. Groups i and iv give no information on which of A and B is the better (though they are very relevant to any inferences about the proportion of cures), whereas each occurrence of ii or iii is a piece of evidence favoring A or B, respectively. On the null hypothesis that A and B have equal rates of cure (which does not contradict the possibility that they might be capable of curing different individuals), the two groups ought to be equally common. Suppose that, of the first n pairs in these two groups, r are in group iii. From mathematical analysis of the problem, we can determine two limits for r (U and L) such that:

a) If r exceeds the upper limit, U, this constitutes significant evidence (at an agreed probability level) against the null hypothesis, and so indicates a higher proportion of cures for B.

b) If r is less than the lower limit, L, this constitutes statistically significant evidence (at the same or a different probability level) against the null hypothesis, and so indicates a higher proportion of cures for A.

c) If r lies between U and L, no decision is yet possible, and the experiment should be continued until results for $(n + 1)$ pairs are available, at which stage the analysis is to be repeated.

The limits U and L depend upon n, and increase as n increases. The smaller the true difference between the rates of cure for A and B, the longer is the experiment likely to continue before one or other of the limits is passed. However, if the difference is very small, its practical importance will be negligible. If a minimum difference that is to be regarded as important can be chosen, significant evidence that the true difference is less than this amount can be adopted as a third rule for terminating the experiment. In this way, the experiment is prevented from continuing indefinitely, and its mean size is much reduced.

Bross has described schemes of this kind, and has shown that the average number of patients required to complete the experiment is of the order of half that required for attaining equal certainty in conclusions when the number of patients to be used is chosen in advance. The advantage is obtainable, of course, only when the experiment is such that the intake of new patients is slow relative to the time that must elapse between treating a patient and obtaining a result. Fisher (1952) has suggested a similar sequential procedure for discriminating between two genotypes by use of the different segregations that their progeny should show. Other uses of similar techniques in biological research will no doubt be found.

7.5. STAIRCASE METHODS

A method in some respects analogous to that of § 7.3 can be used for various estimation problems when only one factor is involved. Suppose that the occurrence or nonoccurrence of a specific response (e.g., death) in animals that have received a particular drug is being studied. Extreme doses will probably produce either response or nonresponse consistently in all animals tested; at any dose in an intermediate range, both responding and nonresponding animals will occur, the relative frequency of response increasing with increasing dose. An important characteristic of the relationship is the *median effective dose* (ED50; cf. § 5.3), or dose just sufficient to cause response in half the animals that receive it; the obvious way of estimating it is to try several doses, to calculate from experiments the proportion of subjects responding at each, thence to derive an equation for the relationship between dose and response rate, and, finally, to find what dose corresponds to 50 per cent response in this equation (Finney, 1952*a*).

If results for individual subjects can be obtained rapidly, a sequential process can be adopted (Dixon and Mood, 1948). A "staircase" of doses can be chosen as any sequence of equally spaced doses (equal spacing on a logarithmic scale being usually preferable). Suitable rules, then, are:

i) Give the first subject a dose guessed to be near the ED50.

ii) If the first subject responds, give the second a dose one step lower.

iii) If the first subject does not respond, give the second a dose one step higher.

iv) Relate the dose for the third subject to that for the second by rules similar to ii and iii, and so continue for all subjects.

These rules concentrate the doses near to the ED50, even though the first dose tested may be a poor guess, and consequently lead to a gain in the precision of estimation. After a preliminary run on a few subjects, it may prove profitable to narrow the interval between steps. Finney (1952*a*, § 55) and Brownlee *et al.* (1953) have discussed the statistical analysis, possible improvements in design, and the merits of the process relative to a nonsequential experiment; Brownlee concludes that in some circumstances it gives a much smaller variance from a specified number of subjects.

Fisher (1952) pointed out that comparisons between feed-

Sequential Experiments

ing programs for animals often need to take account of the most economic levels of feeding and not merely of the responses to arbitrarily selected levels. He proposed to estimate the optimal level of feeding for dairy cattle (and its results) by basing the choice of level in any week on the trend shown in the cost per unit of milk in the previous three weeks, during which, supposedly, three different levels have been tried. Again a fixed staircase of levels could be used, and a set of rules laid down for deciding the level in any week on the evidence of records in the immediately preceding weeks. Extended trial and statistical analysis of variants of this method are needed before their practical utility can be assessed.

CHAPTER VIII

Biological Assay

8.1. Types of Biological Assay

This book is concerned mainly with the general principles of experimental design under the headings of § 1.2. The reader may be interested to see, more fully than has been illustrated earlier, how the principles apply in a particular field; various problems concerned with designing biological assays are discussed below, not merely for their intrinsic importance but to show how these principles can be particularized.

Biological assays are experimental procedures for identifying the constitution or estimating the potency of materials by means of the reactions they produce in living matter. Assays are in regular use in various fields of science, examples being the identification of blood groups by serological tests, the estimation of the potencies of vitamins from their effects on the growth of cultures of microörganisms, and the comparison of insecticides by toxicity tests. Attention is here restricted to *analytical* assays, a particular category that, although of wider application, is of great importance for pharmacological and related purposes. These are experiments to estimate the potency of a *test preparation* (perhaps a natural source of a vitamin) relative to a *standard preparation* containing the same active constituent (perhaps a pure synthetic product). The experimental procedure is to give selected doses of the preparations to subjects, to make on each subject a measurement that is in some way dependent upon the dose, and to use the relationship between this *response* and the dose in order to estimate how much of one preparation is equivalent to one unit of the other. Descriptions of such assays are common in pharmacological literature (Burn *et al.*, 1950); Finney (1951) has given an elementary account. Bliss (1952) and Finney (1952b) have discussed the statistical theory relevant to them, and the account that follows is a brief survey of the ideas on design in this last book.

Analytical assays are such that x units of the test preparation produce the same average response as Rx units of the standard, where R, the *relative potency*, is constant for all x. One important type has the average response, Y, related to dose by the linear *regression equation*

$$Y = a + bx.$$

Here for any particular assay a and b, quantities known as *parameters*, take numerical values such that a is the magnitude of the response associated with zero dose and b is the rate of increase in response per unit increase in dose. This is appropriate, for example, in the assay of riboflavin from its effect on growth of *Lactobacillus helveticus*, the response being a measurement of the acid produced in terms of the titer of sodium hydroxide. If the equation with parameters a and b relates to the standard preparation, that for the test must be

$$Y = a + bRx.$$

The two equations can be shown diagrammatically as two straight lines constrained to intersect at $x = 0$ (Fig. 1). Moreover, the relative potency is the amount of the standard equipotent to one unit of the test preparation, which may be estimated as the ratio of the slopes of the regression equations or the increases in response per unit increase in dose, namely, bR and b. An experiment designed to estimate R in this way

Types of Assay

is termed a *slope ratio* assay. Note that, if the standard preparation has a linear regression equation, the linearity of that for the test and the intersection of the two at $x = 0$ are prerequisites of assayability, for otherwise no single number can express the relative potency.

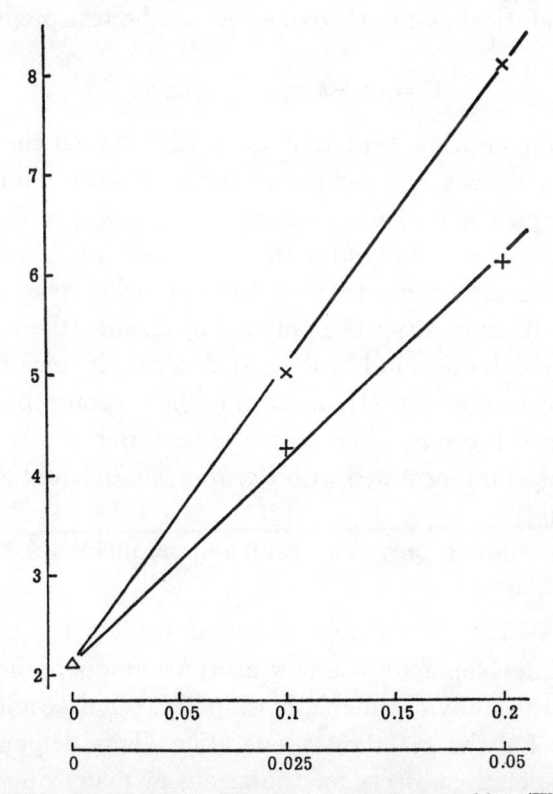

Fig. 1.—Assay of riboflavin in malt, using *L. helveticus* as subject (Wood, 1946). *Upper horizontal scale* (x_S): Dose of riboflavin per tube, in micrograms. *Lower horizontal scale* (x_T): Dose of malt per tube, in grams. *Vertical scale* (y): Titer of $N/10$ sodium hydroxide in milliliters. △: mean response for 4 tubes without treatment; ×: mean responses for 4 tubes on standard preparation; +: mean responses for 4 tubes on test preparation. Two lines intersecting at $x = 0$ have been fitted by standard statistical techniques. The standard line rises by 2.97 ml. per 0.1 μg. riboflavin, the test line by 8.12 ml. per 0.1 gm. malt. Hence the malt is estimated to contain 8.12/2.97, or 2.73 μg. riboflavin per gram.

Even more widely applicable are assay techniques for which the average response is linearly related to the logarithm of the dose:

$$Y = a + b \log x.$$

If this regression equation refers to the standard preparation in an analytical assay, the equation for the test preparation must be

$$Y = (a + b \log R) + b \log x.$$

A diagram showing Y plotted against $\log x$ then consists of two parallel lines, the vertical distance between them being $b \log R$ and the horizontal distance $\log R$ (Fig. 2). *Parallel line assays*, designed to estimate R, the relative potency, from the horizontal distance between two parallel regression lines, are used in estimating the potency of insulin (the response being the reduction in blood sugar of a rabbit injected with a dose of insulin), of streptomycin (the response being the diameter of the zone of inhibition of bacterial growth on the surface of agar inoculated with *Bacillus subtilis*), and of many other drugs.

In this chapter, only slope ratio and parallel line assays are discussed.

8.2. THE STANDARD RESPONSE CURVE

In the development of a new assay technique, a first step must be the study of the relationship between dose and mean response for the standard preparation. This demands the trial of enough subjects for the means at many doses to be estimated with good precision. The response curve need not be linear with respect to dose or log dose, but these two common and important cases illustrate the main ideas adequately. No linear equation can apply for every possible dose, and curvature always appears at extremes.

A simple method of conducting assays against a particular

standard preparation would apparently be initially to determine the response curve for the standard with great care, and thereafter to regard it as a calibration of responses in terms of dose. A batch of subjects could then be given a single dose of a test preparation, the mean response calculated, and the

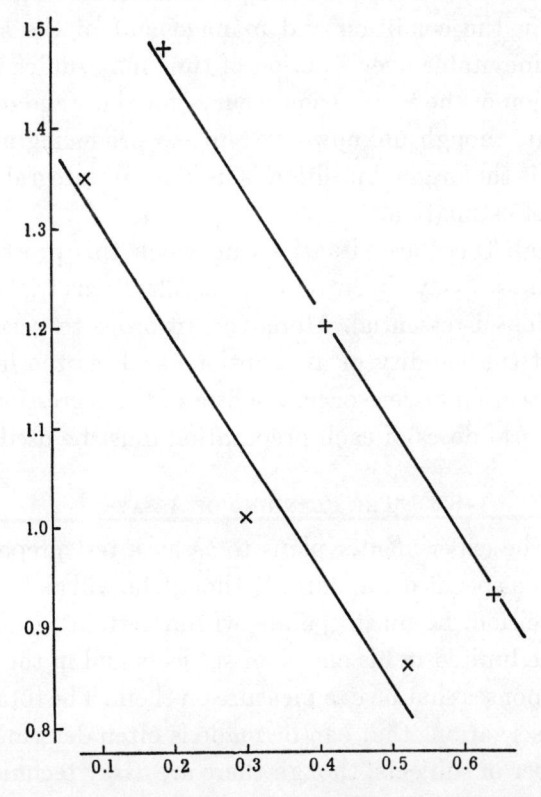

Fig. 2.—Assay of vitamin D in an oil by chick method (Gridgeman, 1951). *Horizontal scale* (x): log daily dose per chick, in units vitamin D or milligrams oil. *Vertical scale* (y): log tarsal-metatarsal distance, in 0.01 mm. ×: mean responses for 28 chicks on standard preparation; +: mean responses for 28 chicks on test preparation. Two parallel lines have been fitted by standard statistical techniques. Measurement shows that the x values of the test line would have to be reduced by 0.224 in order to superimpose it on the standard line. Hence the oil is estimated to contain 0.597 units vitamin D per milligram (since antilog $\bar{1}.776 = 0.597$).

dose of the standard leading to an equal mean response read from the curve; the ratio of doses would estimate the relative potency. Unfortunately, the subjects used for the test preparation cannot be confidently asserted to be perfectly comparable with those used previously for the standard unless they are a sample from the same population. Even the minor changes in the condition and management of the subjects that are inevitable over a period of time may suffice to alter the position of the true response curve for the standard to an important, though unknown, extent, so producing a biased estimate if the original position is used as an integral part of the rule of estimation.

Although there are situations in which this procedure is safe, for most assays in current use simultaneous trial of both preparations is essential. Moreover, in order to permit the testing of the validity of assumptions such as the linearity and intersection at zero or parallelism of the regression equations, several doses of each preparation must be used.

8.3. The Planning of Assays

When the experimenter plans to assay a test preparation, T, against a specified standard, S, though he will aim at maximum precision, he must operate within certain restrictions. He will be limited in his choice of subjects and in the nature of the responses that he can measure on them. The total number of observations that can be made is often determined by the number of subjects, though there are assay techniques in which each subject can be used several times, thus allowing measurement of responses at different doses. Questions on which statistical science is helpful are:

i) What subjects (animals, pieces of animal tissue, microorganisms, etc.) shall be chosen, and what measurement of them shall be used as the response?

ii) What doses of S and T shall be tested, and how many subjects (possibly from a fixed total of N) shall be assigned to each?

iii) How shall doses be allocated to subjects?

Before the statistician can assist with these, he needs an understanding of the experimental problem and knowledge of specific details; his statistical argument needs information from previous similar assays if its conclusions are to be trustworthy. Here the three questions are discussed in reverse order, since that enables their interdependence to be shown more clearly.

8.4. PARALLEL LINE ASSAYS

In the conduct of assays, many of the problems of controlling variability by means of blocks (chaps. iv–vi) arise again; they are briefly reconsidered here in the particular context of bioassay. In view of what has been said in § 8.2, the minimal requirement for a parallel line assay must usually be two doses of each preparation, S_1, S_2 and T_1, T_2, respectively. To have the number of subjects the same at each dose and the two doses of the test preparation in the same ratio as those of the standard, so that the logarithmic intervals are equal, is theoretically advantageous as well as practically convenient.

These widely used *4-point assays* are often arranged as randomized blocks: for example, oestrone has been assayed by taking litters of four female rats and assigning one rat at random from each litter (block) to the four doses, the response being the weight of the uterus after a period of dosing. The cylinder-plate technique used in the assay of antibiotics is often a 4-point assay in randomized blocks, the scheme of experiment being that described at the end of § 8.1. Brownlee *et al.* (1949) have used 8 × 8 Latin squares in microbiological assays of antibiotics, thus accommodating two doses of the standard and two of each of three test preparations for simultaneous estimation of three potencies. The square is used in much the same way as in agricultural trials: the plots are unit inocula of microörganisms, arranged for incubation in a square formation on a growth medium, to which doses of an antibiotic are added, and the Latin square permits the elimination of major positional effects.

Circumstances arise in which blocks of four are not available. Preparations of plant viruses can be assayed by taking single leaves as blocks and inoculating the right and left halves with different doses. A balanced incomplete block design could be used, by assigning to a set of six leaves the six possible pairs of doses from S_1, S_2, T_1, T_2 (with random allocation to the two halves of a leaf) and repeating on further sets of six leaves, but this is not always the best. The four doses can be formally identified with the four combinations of a 2^2 factorial scheme:

1	a	b	ab
S_1	T_1	S_2	T_2

The main effect of A then corresponds to the mean difference in response between the two preparations. The main effect of B is the mean difference in response between the two higher doses and the two lower. These two effects are required in estimating relative potency: their ratio is an estimate of the increase in log dose required to make the doses of the standard equipotent with those of the test preparation; therefore the sum of the ratio and the difference between the logarithms of the doses S_1 and T_1 is an estimate of the logarithm of R. The AB interaction is the difference between the quantities "mean response to S_2 minus mean response to S_1" and "mean response to T_2 minus mean response to T_1"; hence, if the two preparations have parallel lines as their response curves on log dose, the interaction should be zero within the limits of experimental error, and a test of significance of AB is a test of the evidence against parallelism. Provided that the experimenter is confident that the lines really are parallel, he may be willing to sacrifice information on this interaction in order to increase the precision of his estimate of R. He will then confound AB, or, in the present notation, assign doses S_1 and T_2 to some leaves and S_2 and T_1 to an equal number (cf. § 6.10). For his work on southern bean mosaic and other viruses, Price (1946) has proposed such designs as an improvement on earlier experiments (Spencer and Price, 1943) in which B was confounded and the two doses on a leaf were either S_1 and T_1 or S_2 and T_2.

Unless previous experience of an assay technique gives very strong reasons for believing that the assumptions of linearity and parallelism are correct, 4-point assays provide inadequate evidence for testing conditions that are essential to the validity of the analysis. A better choice is the 6-point, using doses S_1, S_2, S_3 of the standard and T_1, T_2, T_3 of the test preparation; successive doses are in a fixed ratio,[1] and equal numbers of subjects are used at all doses (Fig. 2).

This may be likened to a 3 × 2 factorial experiment, in which the main effect of one factor and 1 d.f. from the main effect of the other are used to estimate R; the remaining 1 d.f. from the main effect provides a significance test for deviations from linearity, while the interactions provide other validity tests relating to parallelism. Essentially the same types of design can be used, but, of course, more complex patterns of confounding may be needed. For example, in an antibiotic assay by the cylinder-plate method, the accommo-

1. If S_2 is 1.6 times S_1, then S_3 is 1.6 times S_2 and T_1, T_2, T_3 are in the same ratios.

dation of more than four doses on one plate might be difficult. If sets of three plates had the doses (in random order)

$$\text{I: } S_1, S_2, T_2, T_3$$
$$\text{II: } S_2, S_3, T_1, T_2$$
$$\text{III: } S_1, S_3, T_1, T_3$$

the two most important degrees of freedom would be unconfounded, whereas the validity tests are partially confounded.

With some assay techniques, each subject can be used more than once; after one dose, an interval for recovery is allowed and another dose is applied. For a satisfactory assay, each response must be independent of the previous dosing of the subject. The extreme situation is that in which many tests can be made in fairly rapid succession, so that one or more replicates of all doses can be assigned to the one subject. For example, in the assay of histamine, the contraction of an isolated strip of guinea-pig's gut immersed in a water bath to which a dose is added can be used as a response. With repeated use of one strip of gut, trends in responsiveness may occur, and sets of successive doses can be made into randomized blocks so as to permit the elimination of the major component of trend. Schild (1942) has suggested this and also the further refinement of ordering the sets of doses in accordance with the rows of a Latin square: in a 4-point assay, one piece of gut might be used to give responses to 16 doses, the order of S_1, S_2, T_1, T_2 being taken from successive rows in the second square of Plan 4.6:

$S_1, T_2, T_1, S_2;$ $\quad S_2, T_1, T_2, S_1;$ $\quad T_1, S_2, S_1, T_2;$ $\quad T_2, S_1, S_2, T_1$.

This scheme could be very useful if there were a steady deterioration of responsiveness, as it permits the elimination both of the trend *between* blocks of 4 and of the average trend *within* blocks.

If determination of many responses on each subject is impossible or impracticable, a *cross-over design* provides a valuable compromise. In the rabbit blood-sugar method for insulin assay (§ 8.1), each rabbit can be used more than once, but several days must be allowed for recovery and return to normality after each dose. To test every dose of an assay even once on each rabbit might take too long, and Plan 8.1 shows a possible alternative for a 4-point scheme. The validity test —the interaction between the preparations difference and the levels difference—is confounded between rabbits, but the two

PLAN 8.1

A 4-POINT CROSS-OVER ASSAY FOR INSULIN
(To Be Repeated on Sets of 4 Rabbits)

Dose on Occasion No.	Rabbit No.			
	I	II	III	IV
1.......	S_1	S_2	T_1	T_2
2.......	T_2	T_1	S_2	S_1

main effects are estimated independently of variations between rabbits or between occasions by virtue of the balance in the design. In one assay of this type (Finney, 1952b, § 10.4), 12 rabbits gave a potency estimate as precise as could have been obtained from 132 with only one dose each. So great an increase in precision may more than compensate for the longer duration of the experiment.

Plan 8.1 suffers from the inevitable fault of 4-point assays, inadequate validity tests. If a 6-point scheme of doses can be used, the first two occasions listed in Plan 8.2 will be a great improvement. If completion of the assay can be deferred until each rabbit has been used four times, a still better design can be based upon the three sets of four doses mentioned earlier, each of which occurs for two rabbits (in different order) in the full version of Plan 8.2.

8.5. Choice of Doses for Parallel Line Assays

As in other fields of experimentation, the allocation of doses to subjects is the aspect of bioassay to which statisticians have given most attention. The choice of doses, which precedes this stage, is at least as important to a successful assay. The cost of an assay to the experimenter in terms of time and materials is often roughly proportional to N, the total number of subjects used or responses measured (the number of

PLAN 8.2

A 6-POINT CROSS-OVER ASSAY FOR INSULIN
(To Be Repeated on Sets of 6 Rabbits)

Dose on Occasion No.*	\multicolumn{6}{c}{Rabbit No.}
	I	II	III	IV	V	VI
1	S_1	S_2	S_3	T_1	T_2	T_3
2	T_3	T_2	T_1	S_3	S_2	S_1
3	S_2	S_3	S_1	T_2	T_3	T_1
4	T_2	T_1	T_3	S_2	S_1	S_3

* The first two occasions can be used alone for an assay in a shorter time.

plots). His need, therefore, is to plan for maximum precision in his potency estimate, keeping N fixed and making any necessary provision for testing the validity of assumptions.

Examination of the variance of the estimate indicates that, if an assay could be designed perfectly in other respects and if individual responses to a dose varied little relative to the changes associated with increase in dose, the number of doses of each preparation would not affect the precision of an assay. Such perfection is not attainable, and the effect of number of doses on precision depends upon the closeness with which it can be approached. In order to minimize the variance, the following steps should be taken:

Choice of Doses

i) Choose two doses of the test preparation that are as far apart as possible without appreciable risk of falling outside the range of the linear relationship.

ii) On the basis of any information or intelligent guess about the potency, choose two doses of the standard preparation that are expected to be as potent as the test doses (and are therefore in the same ratio).

iii) For a 4-point assay, use these doses; for a 6-point, 8-point, ..., place 1, 2, ... additional doses of each preparation at regular logarithmic spacing between the extremes.

iv) Divide the subjects equally between all doses.

Steps i and ii presuppose some knowledge about the preparations. If this knowledge is reasonably trustworthy, a good and precise assay can be designed; if not, the assay may have to be only a pilot experiment whose results enable a better one to be planned—a common situation in all experimentation. If linearity and parallelism can be guaranteed, the 4-point design will be the best. If not, a 6-point or 8-point should be chosen, so that tests of validity can be made; the price paid for this, though negligible if almost equipotent doses have been used and the variance of responses is small, may easily be a 10–30 per cent increase in the effective variance of the potency estimate. Serious failure to select equipotent doses, or high response variance and use of only a small number of subjects, can make this loss still heavier. The position is aggravated by an increase in variance per response consequent upon an increased block size made necessary by the larger number of doses.

The importance of distinguishing between study of the response curve, for which many doses are essential (§ 8.2), and conducting an assay should now be clear. Use of more than four doses of each preparation is liable to reduce assay precision seriously and should therefore be avoided unless the response curve is known to be very unstable, because an excessive proportion of the total effort is expended in collecting information on the shape of the response curve.

8.6. Slope Ratio Assays

When the response is linearly related to dose, a *3-point assay* using zero dose and one nonzero dose of each preparation is in some respects analogous to the 4-point for parallel lines, since responses to zero dose estimate a point on both lines. Whatever the responses are, the two regression equations can be made to agree perfectly with the experimental mean responses, so permitting no examination of deviations from linearity or of whether, despite being linear, the true equations for the two preparations fail to intersect at zero dose.[2] The simplest way of providing for such validity tests is by a *5-point assay* (Fig. 1), using one extra dose of each preparation; the two doses of a preparation should be in the ratio 1:2.

Again, randomized block and Latin square designs are useful. If the size of the block is less than the total number of doses, however, the experiment cannot so easily be arranged to confound unimportant comparisons between blocks. Balanced incomplete block designs can, of course, be used, and some gain may result from abandoning balance in favor of a set of blocks that gives greater precision on the most interesting comparisons, less on those wanted only for the less important validity tests. For example, a 9-point design could be put in balanced incomplete blocks of 3 by using 12 blocks (§ 5.5); instead, attention might be concentrated on the slopes of the two lines by using equal numbers of the follow-

---

[2]. This requirement corresponds to that of parallelism and is essential to the validity of the assay procedure.

ing block types (C is zero dose, S_1, S_2, S_3, S_4 and T_1, T_2, T_3, T_4 are doses of the two preparations in the ratio 1:2:3:4):

I: C, S_4, T_4
II: S_1, S_3, T_2
III: S_2, T_1, T_3

This is a particular form of partial confounding. If one subject can be used several times (cf. § 8.4), cross-over designs can be based upon Youden squares; for example, Plan 5.4 could be adapted to a 7-point design, plants being replaced by subjects, leaf positions by three successive uses of one subject, and the letters A–G by the seven doses.

8.7. Choice of Doses for Slope Ratio Assays

Guiding principles for choosing the doses can be developed from consideration of variances (cf. § 8.5). A practical procedure is as follows:

i) Take the highest dose of the test preparation for which there appears to be no risk of its falling outside the region of linearity.

ii) On the basis of any existing information, take a dose of the standard preparation expected to be equipotent.

iii) For a 5-point, 7-point, . . . assay, take zero, these two, and 1, 2, . . . additional doses for each preparation equally spaced between the extremes.

iv) Divide the subjects equally between all doses.

Again some initial knowledge is presupposed, and, in general terms, the remarks of § 8.5 apply. The price that must be paid for validity tests is even greater than with parallel lines, however. Even under the best possible conditions of low variance per response and a successful guess at equipotent doses, a 5-point assay leads to a potency estimate whose variance is 33 per cent greater than for a 3-point with the same total number of subjects, and a 7-point gives a 50 per cent increase. There is no escape from this unless certainty of linearity and of intersection of response lines at zero dose justifies the use of a 3-point: to have an estimate accompanied by adequate validity tests is better than to have an apparently more precise estimate that might in reality be invalid and irrelevant. Although an extravagant number of doses is undesirable in routine assays, experimenters should hesitate to assume that in *their* assays—though perhaps in no one else's—a check on validity is unnecessary!

8.8. Quantal Responses

One type of response frequently used in biological assay is the *quantal* or "all-or-nothing," in which each subject is classified merely as responding or not. Thus a natural way of assessing the potency of insecticides is to try various doses on different batches of insects and to record for each dose how many die and how many survive; an alternative to the blood-sugar technique for insulin assay is to record the occurrence or nonoccurrence of convulsions in mice receiving various doses. These measures of response require special statistical methods for analysis, since they are counts rather than measurements on a continuous scale. However, if at each dose the percentage of subjects showing the response is calculated, a mathematical transformation (Finney, 1952a) can be applied to the percentages in order to give a new measure of response having a linear relation to the logarithm of the dose. Many of the ideas of parallel line assays can then be applied, although there are additional complications in analysis.[3]

Examination of the precision of these assays indicates an interesting new feature: no longer is it desirable to have the

---

3. In some circumstances, the methods of § 7.5 can be applied to estimate the median effective dose for each preparation, the ratio of the two being the potency estimate.

extremes of dose as far apart as possible, and, indeed, precision is much reduced if doses are chosen that give very high or very low percentage responses. Moreover, the ideal spacing of the doses depends in a rather complex manner on the number of subjects used. For example, under certain assumptions about the occurrence of responses, a 4-point assay using a total of 48 subjects will be most precise if the doses can be guessed to give about 20 and 80 per cent responses, whereas if the total number of subjects is increased to 240, the ideal responses rates are about 30 and 70 per cent. As for ordinary parallel line assays, 6-point assays are usually to be preferred, the optimal doses then being those that give about 15, 50, and 85 per cent responses if only 48 subjects are used or, if 240 subjects are used, about 25, 50, and 75 per cent responses. Again the precision of the assay depends in no small degree upon the success with which the dose that will give specified responses can be guessed in advance. Misplaced optimism will have grave consequences if doses believed to correspond to 20 and 80 per cent correspond, in fact, to 2 and 98 per cent, and cautious use of more doses is preferable in cases of doubt.

Any responses measured upon a continuous scale (as in §§ 8.1–8.7) can, of course, be converted to a quantal system by classification as "above" or "below" some arbitrarily chosen level (cf. Table 3.2). This would seriously reduce precision, as well as increase the complexity of the calculations.

8.9. The Choice of Subjects and of Responses

There are often theoretical reasons for believing that the relative potency of two preparations is independent of the species of subject and of the nature of the response measured. This should not be assumed true without good cause: the determination of a relative potency with the aid of mice carries no guarantee that the preparations will have the same relative value in man. In so far as the assumption is justifiable, however, the experimenter may be able to choose between subjects from different sources or between alternative measures of response. As mentioned in § 8.3, this choice is the first concern for a statistician advising on the planning of an assay. Other things being equal, he will prefer subjects that show rapid increase in response as dose increases and little variation in responses at a particular dose. Indeed, for parallel line assays, if past evidence from alternative subjects and types of response used in assaying preparations of the same kind is available, the alternatives can be compared in terms of the quantity

$$\frac{s^2}{Nb^2},$$

where s^2 is the variance of responses at fixed dose, N is the total number of subjects, and b is the rate of increase in mean response per unit increase in log dose (§ 8.1). If values of N for the alternatives are chosen to represent experiments of equal cost, that for which s^2/Nb^2 is least will be the most economic. Bliss and Cattell (1943) and Somers (1950) have given examples of such comparisons. Care in the conduct of the experiment, homogeneity of subjects, and the use of suitable block constraints will help to reduce s^2. The extent to which genetic control of stocks can profitably be used to reduce s^2 or to increase b appears to have been little studied (McLaren and Michie, 1954).

For slope ratio assays, similar comparisons are more awkward to make, though a good approximate rule is that of seeking to minimize

$$\frac{s^2}{NB^2},$$

where B is the total increase in response between zero dose and the highest dose on the linear section of the response curve.

General considerations suggest that potency estimates based upon quantal responses will be less precise than estimates from similar experiments using quantitative responses (with the same total number of subjects), though this is not invariably true. On the other hand, quantal response techniques can be used when others cannot and, even when this is not so, may be so much simpler and less costly as to permit many more subjects to be used. If a quantal response is to be used, rapid increase in the percentage of responses with increasing dose is desirable. In assays of the trypanocidal activity of neoarsphenamine, Morrell and Allmark (1941) report slight success in selective breeding of rats for this property. Miller's (1944) account of the comparison of alternative techniques for digitalis assay is a good practical illustration of the principles enunciated here.

CHAPTER IX

The Selection of a Design

9.1. Design, Analysis, and Interpretation

The reader who has understood previous chapters ought by now to be aware of two general principles, although these have not been explicitly stated earlier:

i) The design of an experiment has a great influence on the form of statistical analysis appropriate to the results.

ii) The success of an experiment in answering the questions that interest the experimenter or in pointing to profitable lines for further study, with reasonable economy of time and resources, depends largely upon right choice of design.

In the broad sense these principles are obvious: the form of statistical analysis must depend upon what experiment has been done, and unless an experiment is planned to be relevant to the subject of study, it can scarcely give useful answers! Detailed application of the principles goes much deeper.

The nature of the dependence of analysis on design and on knowledge or assumptions about algebraic models for the behavior of measurements is most conveniently discussed in books on statistical analysis, Kempthorne being perhaps the most detailed in relation to designs described in the present book; a few simple ideas have been mentioned in §§ 3.3, 3.6, 4.8, 4.11, 5.7, 6.7. More fundamental than this statistical technique, though closely related to it, is the choice of a design for an experiment, this comprising decisions under headings i–iv of § 1.2. Too often the statistician's interest in de-

Design, Analysis, Interpretation

sign is thought to be almost confined to heading iii, all other decisions being for the experimenter alone. Unless the experimenter is himself skilled in statistical science, however, he is unlikely to appreciate fully how these decisions are related to the specification of the questions that the experiment is competent to answer and to the reliability of the answers obtainable.

Although written twenty years ago, a paper by Yates (1935) on "complex experiments" contains much sound advice on the relative merits of different designs that is still imperfectly appreciated. A more recent but less weighty paper (Finney, 1953) shows how some of the general principles of this chapter apply to the special field of agricultural research. General papers are necessarily inadequate, and experience of experimentation in a particular branch of research is essential to the making of the best choice of designs.

The experimenter, possibly unaccustomed to involved quantitative reasoning, may not notice how an obvious and simple research program can be much improved by ingenuity of design, either without appreciably increasing the cost[1] or for an increase in cost that is more than compensated by the increase in information. The statistician, attempting to express mathematically the requirements of a biological problem, may oversimplify some aspects and overcomplicate others and so produce impracticable proposals. The sections that follow are concerned primarily to illustrate headings i, ii, and iv of § 1.2, subjects that cannot be formalized as readily as those arising from iii and that need wide knowledge of the particular field of application for their complete elaboration. Here the statistical point of view is stressed, but with full

[1]. Throughout this chapter, *cost* is to be regarded as referring to expenditure of money, materials, labor, time, or any other factor limiting the extent of the research (cf. § 1.2).

The Selection of a Design

recognition that collaboration and not dogmatic assertion is required from the statistician: where compromise on optimal statistical considerations is found inevitable, the experimenter will no doubt have the last word, but the statistician's duty is to inform him of the probable consequences. Discussion between experimenter and statistician can lead to a complete change in the character of an experiment, not because of insistence by the statistician but because both come to realize more clearly the issues involved and the best way of exploiting the principles of design for the purpose. What follows is to be regarded as illustrative of important lines of argument rather than as a comprehensive account.

Too many experiments are undertaken in a spirit of "Let us try a miscellaneous set of alternatives, measure anything that looks interesting, and see whether any important differences emerge." Such experiments may be valuable in the preliminary investigation of a new field, where they are useful as providing pointers to profitable lines of detailed research rather than as themselves giving exact results. Their excessive use derives from inadequate consideration of research strategy and unwillingness to direct attention to problems that are both important and likely to yield to attack with the methods and resources available. Unless an experiment is planned so that the treatments tested and their scheme of allocation to subjects are directed to the answering of specific questions, the most important results are unlikely to be achieved. In a research organization the experienced statistician should be able to persuade his colleagues to specify the major objectives of an experiment,[2] to exclude trivial or irrelevant topics, and to employ a design especially suitable for the purpose rather than a casual assembly of treatments. Al-

---

2. Often this includes the need for more exact definition of the usage of such words as "best," "larger," "efficient," "fertility," "environment."

though this practice can result in an experiment's becoming larger and more complex than was originally intended, the statistician must beware of urging that experiments be made unnecessarily elaborate; limitation of resources (human or material) or restriction of interest sometimes makes a very simple design preferable to one that is formally more efficient. On the other hand, he must be prepared occasionally to express a firm opinion that, unless an experiment can be expanded considerably, its chances of answering any of the questions put to it are so slender that it might as well be abandoned. Such assistance to research is of far greater value than the performance of routine computations: a well-designed experiment will usually allow its conclusions to be easily obtained, whereas no computations, however industriously or ingeniously performed, can produce entirely satisfactory conclusions from an ill-designed one. Considerable tact is needed in discussion of these matters; unless the experimenter has previously benefited from similar assistance, he is apt to distrust or resent criticism of his choice of treatments, of the number of levels of a factor, or even of the whole concept of his experiment by one who is not a specialist in the same field of science.

9.2. The Number of Factors

Factors additional to those for the study of which an experiment was first contemplated can often be incorporated, without appreciable loss to any aspect of the original study. This is particularly likely if the precision required for the original purpose demands extensive replication, for replication in respect of the first set of factors and their interactions is not lowered by the inclusion of others factorially (chap. vi). Even small experiments, however, allow opportunities of this kind. If a 2^3 experiment is wanted, anything smaller than

The Selection of a Design

four randomized blocks of eight would rarely give adequate replication; as shown by Plan 6.2, two additional factors can be included with very slight loss to the original experiment (a reduction in d.f. for error) and with great gain in respect of information on the new treatments and the interactions with the others. Many workers with 3^3 designs use four replicates in blocks of nine, with the mistaken idea that balancing the confounding (§ 6.11) has special advantages; three replicates would often give adequate precision on these three factors, and the design then has the merit of permitting the introduction of one additional factor, or even of two by having one-third of a replicate (§§ 6.9, 6.10).

Factors additional to the first intention may be incorporated into the design at the beginning, or the desirability of further differentiation in the treatment of the plots may appear later. Thus, even if an experiment is not "saturated" with factors initially, there are advantages in choosing a confounding system that will permit later additions in preference to partial confounding. No condemnation of experiments with several replicates of every combination of treatments is intended. In many situations, however, saturation with factors so as to give one replicate, or even "supersaturation" in the form of fractional replication, enables the experimental labor and materials to be used more advantageously. Many experimenters fail to consider whether other factors relevant to their subject could not profitably be investigated simultaneously with those for which an experiment was begun.

9.3. The Choice of Levels

For some factors, the number of levels to be tested leaves little choice to the experimenter. His interest may be restricted to the comparison of two or more qualitatively distinct

Choice of Levels

states: male and female; apparatus of three alternative forms; five different strains of bacteria. Nevertheless, in researches that involve a fairly large number of treatments without any factorial structure among them, as, for example, the plant breeder's tests of new varieties or the entomologist's comparisons between insecticides, the exact number to be included in any one experiment is often not rigidly specified. Probably incomplete block designs of some type (chap. v) will be wanted, and the addition or omission of one or two treatments may greatly help the selection of a design. An experiment on 8 treatments that had to be conducted in blocks of 3 could be arranged in balanced incomplete blocks only by having a minimum of 168 plots (21 replicates). As an alternative to an awkward, partially balanced design, the addition of one more treatment would make possible lattice and lattice square designs, those in 4, 8, 12, . . . replicates being balanced. If one treatment could be omitted, a design in 3 or any multiple of 3 replicates is possible (Plan 5.2). Moreover, unless conditions imposed by the experimenter or his materials rigidly determine the number of plots per block, the possibility of slight alteration from the size of block first proposed makes the choice of design freer and avoidance of designs with little balance easier.

If the number of treatments cannot be changed from one that is awkward for design, inclusion of one or two of the more interesting treatments with double replication may be helpful. Such a treatment would be regarded formally as two distinct treatments throughout the construction and analysis of the experiment, but duplicate results would finally be averaged (§ 9.4).

Often the levels of a factor are arbitrary values on a continuous scale: the amount of Epsom salt to be used in a growth medium for *Drosophila;* the temperature of an incu-

The Selection of a Design

bator; the date on which seeds are sown. The experimenter is not usually concerned solely with the levels tested in his experiment but wishes to make inferences about other levels. He may want a general idea of the shape of the curve relating the average value of a measurement to the level of a factor, or he may be interested in some more restricted aspect, such as the level at which his measurement assumes its maximum value or is optimal in the sense of showing the greatest net profit after allowance for the cost of treatments applied.[3] Unless he is confident that the relationship is linear (so that no maximum exists) within his range of interest, he needs at least three levels. For estimating the level giving maximal or optimal returns, the ideal number of levels depends largely on the reliability of existing information on the quantity sought: if a reasonably good prediction can be made in advance, three will suffice; if not, the problem involves a fairly thorough study of the curve and four or five are wanted (cf. § 7.3).

The practicability of confounding schemes and the flexibility of designs for including many factors are greatly aided by having all factors at the same number of levels, although factors at 2 and 4 levels can be mixed satisfactorily. Hence 2^n and 3^n designs are of greatest importance; 4^n and 5^n are equally sound in theory, but, despite confounding and fractional replication, the large number of treatment combinations limits their practical use. Mixed designs (§ 6.4) should be avoided unless there are particularly strong reasons for having factors with different numbers of levels.

When levels are measured quantitatively, it is usually desirable to have equal intervals between successive values. If

---

3. This last, of course, is especially important in applied science, such as studies of the fertilizer needs of crops or the materials and physical conditions needed for a penicillin factory.

theory indicates an approximate linear dependence on the logarithm of the level rather than on its absolute value, equal logarithmic spacing is better; this consideration often reinforces the practical convenience of testing a series of dilutions in geometric progression: 1/10, 1/100, 1/1,000, or 1/4, 1/8, 1/16, 1/32. In an experiment intended for the study of a maximum or an optimal, the middle level tested should correspond approximately to any a priori knowledge or guess about the maximum. A common fault in experiments for comparing different materials with similar modes of action (different phosphatic fertilizers or different sources of a vitamin) is to use levels so high that all materials supply adequate amounts of their important constituents and no differences appear, or levels so low that responses of any kind can scarcely be detected. Detailed recommendations on the choice of levels depend upon knowledge of the type of relationship between level and effect and upon the purpose of the experiment; §§ 8.5, 8.7, and 8.8 provide illustrations, and §§ 7.3 and 7.5 discuss other special problems.

9.4. Controls

An experimenter sometimes argues that he knows a certain type of treatment to have beneficial effects and that he is interested only in comparing alternative forms of it. Nevertheless, unless he is certain—a rare state of mind—that benefit occurs in all circumstances, he should include plots without this treatment, or *controls*. For example, a hormone might be known to affect the growth of plants, but an experiment in which two different methods of application were compared might be very misleading unless it included untreated plants: the absence of any clear difference between the two treatments could mean either that the two were equally effective or that special circumstances had prevented the plants from

The Selection of a Design

responding to the hormone, and only comparison with controls can distinguish between the explanations.

This can be particularly important in clinical medicine, especially if faith in a remedy may effect a cure. A good example has been described in § 2.10. Ethical considerations sometimes prevent the inclusion of true controls in a clinical trial. Neither statistician nor experimenter can escape this restriction, but both have an obligation to search for an experimental procedure that is both ethical and free from logical difficulties in interpretation.

When an experiment is designed to compare each of a large number of treatments with a control, additional replication of the control is desirable. For example, a biochemist might wish to compare many alternative diets with a standard, in terms of their effect on rat metabolism, or a plant breeder might wish to test a series of new strains of a cereal for their yields relative to that of a variety in current use. Maximum precision for a fixed total number of plots is then achieved by allocating more plots to the control than to any one other treatment. The ideal is that the ratio of numbers of plots should be the square root of the number of other treatments, but in practice the integers nearest on either side of this square root (i.e., 4 or 5 if there are 21 treatments) will give almost optimal results. Hence the appropriate practice is to include the control as though it were several distinct treatments, to design and analyze the experiment accordingly, and, finally, to average the means obtained for these quasi-treatments (§ 9.3).

9.5. Number of Replications

Sometimes the best service that a statistician can render to an experimenter is to tell him that, unless he can substantially increase the number of replications in a proposed experiment, he has little hope of obtaining for the comparisons that interest him a standard error small enough to make the results useful. If a larger experiment is impossible, the experimenter should turn his attention to something different rather than squander his resources on efforts unlikely to give any return. Too often, small experiments on three or four treatments in one or two replicates are conducted under the guise of "observation plots" or "demonstration trials."[4] No one would deny the value of preliminary observations on a few experimental units as a guide to future lines of research, or of demonstrating established results so as to educate others; the fault lies in the use of these names when enough is known for casual observations to be no substitute for precise experiments but not enough is known for further tests to be regarded merely as demonstrations of accepted truths to a wider public (students, farmers, etc.).

Sometimes a statistician may be able to state that a proposed experiment gives more replication than is needed. That this occurs less often is perhaps attributable to the eternal optimism of experimenters rather than to the excessive demands of statisticians!

The ideal number of replicates depends upon consideration of standard errors in relation to costs and to the magnitudes of effects that are of interest. Inevitably, compromises are needed, and recommendations for any experiment can be based only on indications from similar work in the past. Cochran and Cox (1950) have given a valuable discussion. If the variance per plot (§ 3.3) can be guessed in advance, from the evidence of previous experiments, as s^2, and a difference between two treatments is required to have a standard error of e, then the number of plots of each treatment should be $2s^2/e^2$ or the next larger integer. Unsuccessful guessing of s^2 will make the standard error actually achieved greater or

[4]. These euphemisms are also often made the excuse for lack of randomization.

The Selection of a Design

less than e, and, in order to reduce the risk of exceeding e, this number of replicates must be increased. Rules can be developed either by raising the probability that the standard error will not exceed e or by specifying a probability that, if the difference between the means exceeds an arbitrary amount, it will be detected as statistically significant.

Sequential design (chap. vii) is another way of achieving a predetermined precision or sensitivity in an experiment: in theory it does this more exactly than the schemes just described, but in practice there are many problems in which sequential application of treatments is impossible.

9.6. Allocation of Treatments to Plots

Not until provisional decisions have been taken on the issues discussed in previous sections, should questions relating to blocks, confounding, restrictions on randomization, and the like, be answered, although they can usefully be kept in mind from the start. The general character of the design has been fixed by choice of the number of factors, the number of levels, and the number of replications; specifications on replication, however, are rarely absolutely rigid, and a slight increase or decrease in the number of replicates is usually permissible in response to other needs of design. Indeed, at this stage the absolute impossibility of complying with certain specifications must be remembered. Not even a committee of statisticians can devise a Graeco-Latin 6×6 square, a 2^5 design in blocks of 4 with all main effects and 2-factor interactions unconfounded, or a balanced incomplete block design for 6 treatments in blocks of 4 in less than 10 replicates!

The maximum number of plots per block is often fixed by the nature of the experimental material (§ 4.7). Even if no absolute maximum is set by the number of animals per litter, the number of observations that one worker can be expected to complete in a session, or some analogous consideration, experience is likely to show that, beyond a certain number, increasing heterogeneity of plots within a block more than balances the convenience of including many different treatments. In any field of research, examination of records of past experiments helps to indicate the main sources of variation and the consequences of using different sizes of block or of blocks defined by alternative characteristics of the plots. If the total number of treatments is small, randomized blocks or Latin squares will usually be the preferred arrangements. If the number is larger than the desired block size and no factorial structure is present, balanced incomplete blocks or Youden squares will be the aim, with a lattice or other partially balanced design as the escape from excessive replication. If the many treatments are combinations of several factors, confounding of interactions will usually be the best way of limiting block size, and fractional replication may provide a way of studying many factors in one experiment. The possibility of modifying block sizes or numbers of factors and levels so as to form a good design must often be considered. For example, a 3×2^2 or $3^2 \times 2$ factorial scheme needs at least 36 plots for satisfactory balance (Plan 6.5); if blocks of 9 can replace blocks of 6, a 3^3 in 27 plots offers many advantages, despite its smaller size (§ 9.3).

Having decided what sets of treatments are to be assigned to the various blocks, it only remains to insure that the order within a block is randomized and that, if incomplete blocks are used, the order of allocation of the sets to the blocks of experimental material is also random (§§ 3.2, 4.5, 5.7).

9.7. THE NUMBER OF EXPERIMENTS

In an applied science such as agriculture, expenditure on research must be largely governed by the probable gain from

The Selection of a Design

use of the results. In pure research, as emphasized in § 1.2, economic limitations may be less apparent, but, in the last analysis, the amount of experimentation undertaken on any topic is determined by the value of the results to the general progress of science. Questions relating to the amount of experimentation that should be undertaken are discussed here in terms of research directed to a practical objective, where the ideas can more readily be expressed quantitatively, but they are not entirely irrelevant to pure research.

Despite the precautions taken in the conduct of an experiment, the conclusions obtainable may not be representative of all conditions under which results are wanted; the precision estimated from internal evidence will then exaggerate the consistency that would be shown if any particular numerical comparison were repeated by different investigators or under different conditions. The response of a crop to a fertilizer will depend upon soil, upon seasonal factors, and upon the general management of the crop; the effect of dietary supplements upon animal growth will depend upon the basal diet and normal management of the animals, as well as upon their genetic constitution, age, and past history. If the average effect of a proposed change in crop or animal husbandry is to be assessed precisely, experiments of similar type must be widely distributed over the whole region or population to which the results are eventually to be applied. Only if the possibility of differences between treatment effects at different places or on different subjects can be dismissed, will adequate precision be achieved as satisfactorily by one highly replicated experiment.

When a suitable design for a single experiment has been selected, how many of these should be performed (or, if that procedure is to be adopted, by how much must the replication of the one experiment be increased) to satisfy economic considerations? Yates (1952) has discussed this question with reference to estimation of the optimal amount of some material[5] to be recommended for commercial practice. The greater the number of experiments, the more precisely will the optimal be estimated and the smaller will be the loss from recommending an amount that differs slightly from the true most economic level. Against this must be set a total cost of experiments that increases approximately in proportion to their number. After showing that the expected loss from imperfect estimation of the optimal will be approximately proportional to the variance of the estimate, Yates determined the number of experiments that would minimize the total of cost of experimentation and loss by failure to recommend the most profitable level. His result is

$$\left(\frac{kvT}{c}\right)^{1/2};$$

here v is the variance of the estimated optimal amount per unit application (per acre, per animal, etc.) as given by a single experiment, T is the number of units to which the recommendation will be applied, c is the cost per experiment, and k is a constant relating the variance to the value of the expected loss.

This result is not to be interpreted too rigidly, but it does give a basis for assessing the desirable number of experiments on economic grounds, instead of the more usual reliance upon the whim of those who control research funds. More recently, Grundy *et al.* (1954) have developed a method for deciding the number of experiments to be undertaken when the point at issue is which of two alternative practices (in which no quantitative variations are envisaged) ought to be recommended for general adoption; both theory and rule are more

---

5. For example, the amount of fertilizer per acre for a particular crop or the amount of a particular component of animal feeding stuff.

The Selection of a Design

complicated, though for practical purposes the rule can be used by reference to a single table or diagram.

9.8. THE MEASUREMENTS

On each plot of an experiment, some measurement (or count) must be made for use as an assessment of the integrated consequences of all treatments and other conditions pertaining to the plot. The nature of this measurement is obviously decided by the purpose of the experiment: if the experiment has been planned to compare the effects of different diets on the amount of vitamin A in rats' livers, the possibility that length of tail might be a measurement less subject to variation, therefore giving relatively higher precision to comparisons, is irrelevant! Nevertheless, there is often room for some choice in the exact definition of the measurement: How long a time shall elapse between the start of the experiment and the measurement? What mechanical, chemical, or biochemical techniques shall be adopted as the standard procedure for making the measurement? If the measurement is to be on only a sample of the whole "plot,"[6] what size must this sample be and how shall it be selected?

To questions such as these, no general answer can be given. As for decisions on the specifications of the plots, to which, indeed, they are closely related, the best answers are largely empirical and can be discovered only from study of previous experiments and records. When alternative types or sizes of plot or alternative forms of measurement of result are regarded as equally valid and relevant to the subject of investigation, choice between them should depend upon which is likely to give the more precise estimates of treatment effects. Anal-

---

6. For example, a small piece of a particular tissue may be submitted to chemical analysis, reticulocyte counts will be made on only a small sample of blood, and yields or assessments of disease incidence may sometimes be based on only a fraction of all plants in a field plot.

The Measurements

yses of variance of other experiments and examinations of the components of variation according to procedures now familiar to statisticians can be used to predict the relative precision of the alternatives. Once again, the building-up of a corpus of experience in a particular field of research is vital to the improvement of experimental design and practice in that field (cf. § 9.6), though interest in what is essentially a point of experimental technique must not be allowed to delay indefinitely the start of the real research program.

In practice, several entirely different measurements may be wanted from each plot, these relating to different aspects of the effects of treatments. The same principles will guide the choice of each, and the decision on what design shall be adopted will have to represent a compromise between the alternatives that seem most suitable for the various measurements.

9.9. Concomitant Measurements

The precision of comparisons between treatments can sometimes be much increased by judicious use of additional measurements on each plot of one or more concomitant properties of the plot. The aim is to eliminate inherent differences between plots in respect of characteristics present before the treatments were given or known to be unaffected by the treatments. By the technique of *covariance analysis*, the internal evidence of the experiment can be used to estimate the magnitude of the difference between two values of the *dependent variate* (the measurement under study) that is associated with unit difference between corresponding values of one of these *independent variates*. The computations, which are not described or illustrated here, are an extension of the analysis of variance. With the aid of this estimate, all values of the dependent variate can be adjusted to equality in the independent variate, and a corresponding reduction in the

The Selection of a Design

variance is made, in order to allow for the variation thus eliminated.

The precision of the experiment is increased only if the independent variate really is associated with the measurement under study: causation is not essential, but it must in some way be an index of factors that influence the measurement. Moreover, the independent variate must itself be unaffected by the difference in treatments (lest the adjustment described be totally misleading), a requirement usually met by using a variate that is measured before differential treatments are applied. The experimenter will therefore be wise to measure initially any characteristic of his experimental units that might later be useful in this way, at least on the assumption that this requires little extra effort; this advice would be inappropriate if the additional measurements required so much work that they could be obtained only at the price of a serious reduction in the size of the experiment. Pre-treatment records of the quantity to be studied in the experiment may be valuable, but often these are impossible to obtain (e.g., internal measurements of animals), and experience or judgment may suggest some useful alternatives.

Initial weights of animals, yields of plots in a pre-experimental season, blood-sugar values before treatments involving different doses of insulin are given, and similar records form ideal independent variates. The decision whether or not to make the recording of an independent variate part of the design should rest upon previous experience, this being yet one more way in which past results help future planning. Sometimes one or more additional variates are recorded almost inevitably as part of the experimental routine. In other experiments, careful thought has to be given to whether the labor of measuring independent variates could not be more profitably given to increasing the replication of the experiment.[7] Theory and experience indicate that a concomitant measurement made before an experiment begins can usually be employed more advantageously in a covariance analysis than as a basis for grouping the plots into homogeneous blocks; in fact, the possibility of covariance for this variate leaves open the opportunity of constructing blocks by reference to some other qualitative characteristic of the plots (§ 4.7).

An experiment reported by Kodlin (1951) provides an interesting illustration of the value of covariance analysis. Two groups of 10 animals were used in a comparison of the effects of substances A and B on blood pressure. Analysis of the blood pressures at the end of the experiment by the method of § 3.3 showed a difference of 12.5 ± 4.94 mm. Very naturally, the experimenter had also recorded initial blood pressure for each animal. A common procedure would be to compare A and B in terms of reductions from the initial values instead of the final values alone, and this analysis showed a difference of 7.0 ± 5.08 mm. Thus the precision (as indicated by the standard error) was not improved in the least by making this obvious allowance for initial values. Yet there was quite a close correlation between initial and final blood pressures for the same animal; when a covariance analysis was used to adjust the final values to a basis of initial equality, the treatment difference was estimated at 9.8 ± 3.28 mm. In fact, the comparison based upon reductions in blood pressure overcorrected for initial values, whereas the covariance analysis enabled

7. A related point is that, although in an agricultural experiment the yields of a previous crop might be useful in covariance, it is rarely desirable to spend a year in determining these on untreated plots rather than to begin the experiment immediately. If for other reasons pretreatment yields have been measured, they should be tried in a covariance analysis, but otherwise the measurement of a concomitant is likely to be a poor compensation for delay in obtaining measurements on treated plots.

The Selection of a Design

the comparison between A and B to be made as precisely as a comparison based on final values alone from two groups of 23 animals. In the experiment of §§ 3.2 and 3.6, body weight was recorded, but unfortunately not for every rat. Moreover, these weights were taken at the end of the experiment, so that there are dangers in using them in a covariance analysis, though justification may be claimed because the weights themselves appear not to have been affected by the treatment difference. The records available suggest that a covariance on body weight would have increased the efficiency of the paired design by a further 41 per cent (cf. § 3.8).

9.10. Statistical Analysis

For most experiments, the labor of statistical analysis is small relative to the total cost of the experiment. When this is so, the choice of a design should be scarcely influenced by consideration of whether or not the results will be easy to analyze. The symmetry of a well-designed experiment usually insures that the analysis is not excessively laborious, and, for designs of the types discussed in earlier chapters, standard computational procedures are well established. In some circumstances, the conduct of an experiment may be so simple that results can be produced rapidly: inclusion of additional plots in order to allow the use of a balanced incomplete block design instead of a partially balanced, or of complete instead of incomplete blocks, may then save so much time and labor in computation as to outweigh the extra work in the experiment.

The attitude toward this matter of an experimenter who must do his own computing and who has no calculating machine will naturally differ from that of one in an organization with a well-equipped computing section. In any field of biology in which extensive numerical records are obtained, a cal-

Statistical Analysis

culating machine is an investment whose small cost is rapidly repaid by the saving of time, the increase in accuracy, and the practicability of computations previously thought prohibitively laborious that its use makes possible. A machine should be regarded as an indispensable adjunct to quantitative biological research, and an operator especially skilled in its use is an obvious economy if the volume of computing is large. This point is quite distinct from that of employing a statistician, and much research would benefit from the realization that a more systematic approach to its computations need not await the appointment of a statistician. Nevertheless, any biologist who has read this far will realize that he also needs access to the advice of a statistical specialist if he is to make the best use of modern ideas in experimental design.

References

A. Elementary Approach to Statistical Analysis

BERNSTEIN, L., and WEATHERALL, M. 1952. Statistics for medical and other biological students. Edinburgh: E. & S. Livingstone.

FINNEY, D. J. 1953. An introduction to statistical science in agriculture. Copenhagen: Einar Munksgaard.

HILL, A. B. 1950. Principles of medical statistics. 5th ed. London: The Lancet.

MAINLAND, D. 1938. The treatment of clinical and laboratory data. London: Oliver & Boyd.

———. 1952. Elementary medical statistics. Philadelphia: W. B. Saunders Co.

MORONEY, M. J. 1953. Facts from figures. 2d ed. London: Penguin Books.

QUENOUILLE, M. H. 1950. Introductory statistics. London: Butterworth-Springer.

SNEDECOR, G. W. 1946. Statistical methods. 4th ed. Ames: Iowa State College Press.

TIPPETT, L. H. C. 1949. Statistics. 5th impression. London: Oxford University Press.

B. Standard Texts on Experimental Design

COCHRAN, W. G., and COX, G. M. 1950. Experimental designs. New York: John Wiley & Sons.

DAVIES, O. L. (ed.). 1954. The design and analysis of industrial experiments. London: Oliver & Boyd.

FISHER, R. A. 1951. The design of experiments. 6th ed. London: Oliver & Boyd.

FISHER, R. A., and YATES, F. 1953. Statistical tables for biological, agricultural and medical research. 4th ed. London: Oliver & Boyd.

KEMPTHORNE, O. 1952. The design and analysis of experiments. New York: John Wiley & Sons.

KITAGAWA, T., and MITOME, M. 1953. Tables for the design of factorial experiments. Tokyo: Baifukan Co.

References

QUENOUILLE, M. H. 1953. The design and analysis of experiment. London: Charles Griffin & Co.

YATES, F. 1937. The design and analysis of factorial experiments. Harpenden, England: Imperial Bureau of Soil Science.

C. OTHER REFERENCES

BACHARACH, A. L. 1940. The effect of ingested vitamin E (tocopherol) on vitamin A storage in the liver of the albino rat. Quart. J. Pharm. & Pharmacol., 13:138–49.

BACHARACH, A. L., CHANCE, M. R. A., and MIDDLETON, T. R. 1940. The biological assay of testicular diffusing factor. Biochem. J., 34:1464–71.

BIGGS, R., and MACMILLAN, R. L. 1948. The error of the red cell count. J. Clin. Path., 1:288–91.

BLISS, C. I. 1952. The statistics of bioassay. New York: Academic Press.

BLISS, C. I., and CATTELL, McK. 1943. Biological assay. Ann. Rev. Physiol., 5:479–539.

BOX, G. E. P., and WILSON, K. B. 1951. On the experimental attainment of optimum conditions. J. Roy. Statist. Soc., B13:1–45.

BROSS, I. 1952. Sequential medical plans. Biometrics, 8:188–205.

BROWNLEE, K. A., HODGES, J. L., and ROSENBLATT, M. 1953. The up-and-down method with small samples. J. Am. Statist. A., 48:262–77.

BROWNLEE, K. A., LORAINE, P. K., and STEPHENS, J. 1949. The biological assay of penicillin by a modified plate method. J. Gen. Microbiol., 3:347–52.

BURN, J. H., FINNEY, D. J., and GOODWIN, L. G. 1950. Biological standardization. London: Oxford University Press.

CHINLOY, T., INNES, R. F., and FINNEY, D. J. 1953. An example of fractional replication in an experiment on sugar cane manuring. J. Agr. Sc., 43:1–11.

COCHRAN, W. G., AUTREY, K. M., and CANNON, C. Y. 1941. A double change-over design for dairy cattle feeding experiments. J. Dairy Sc., 24:937–51.

COX, G. M., and COCHRAN, W. G. 1946. Designs of greenhouse experiments for statistical analysis. Soil Sc., 62:87–98.

DAVIES, O. L., and HAY, W. A. 1950. The construction and uses of fractional factorial designs in industrial research. Biometrics, 6:233–49.

DIXON, W. J., and MOOD, A. M. 1948. A method for obtaining and analyzing sensitivity data. J. Am. Statist. A., 43:109–26.

EMMENS, C. W. 1948. Principles of biological assay. London: Chapman & Hall.

FABERGÉ, A. C. 1943. Genetics of the Scapiflora section of *Papaver*. II. The alpine poppy. J. Genetics, 45:139–70.

References

FINNEY, D. J. 1947. The construction of confounding arrangements. Empire J. Exper. Agriculture, 15:107–12.

———. 1951. Biological assay. Brit. M. Bull., 7:292–97.

———. 1952a. Probit analysis. 2d ed. London: Cambridge University Press.

———. 1952b. Statistical method in biological assay. London: Charles Griffin & Co.

———. 1953. Response curves and the planning of experiments. Indian J. Agr. Sc., 23:167–86.

FISHER, R. A. 1926. The arrangement of field experiments. J. Ministry Agriculture, 33:503–13.

———. 1942. The theory of confounding in factorial experiments in relation to the theory of groups. Ann. Eugenics, 11:341–53.

———. 1952. Sequential experimentation. Biometrics, 8:183–87.

FLOYD, J. C. 1949. Penicillin formulations: the efficacy of oily injections. J. Pharm. & Pharmacol., 1:747–56.

GRIDGEMAN, N. T. 1951. On the errors of biological assays with graded responses, and their graphical derivation. Biometrics, 7:200–221.

GRUNDY, P. M., REES, D. H., and HEALY, M. J. R. 1954. Decision between two alternatives—how many experiments? Biometrics, 10:317–23.

HARDY, G. H. 1908. Mendelian proportions in a mixed population. Science, 28:49–50.

HARRISON, E., LEES, K. A., and WOOD, F. 1951. The assay of vitamin B_{12}. Part VI. Analyst, 76:696–705.

HERWICK, R. W., WELCH, H., PUTNAM, L. E., and GAMBOA, A. M. 1945. Correlation of the purity of penicillin sodium with intrasmucular irritation in man. J.A.M.A., 127:74–76.

HILL, A. B. 1951. The clinical trial. Brit. M. Bull., 7:278–82.

JELLINEK, E. M. 1946. Clinical tests on comparative effectiveness of analgesic drugs. Biometrics, 2:87–91.

KALMUS, H. 1943. A factorial experiment on the mineral requirements of a *Drosophila* culture. Am. Naturalist, 127:376–80.

KODLIN, D. 1951. An application of the analysis of covariance in pharmacology. Arch. internat. de pharmacodyn. et de thérap., 87:207–11.

LOUDON, I. S. L., PEASE, J. C., and COOKE, A. M. 1953. Anticoagulants in myocardial infarction. Brit. M. J., 1:911–13.

MCLAREN, A., and MICHIE, D. 1954. Are inbred strains suitable for bioassay? Nature, 173:686–87.

MATHER, K., BOWLER, R. G., CROOKE, A. C., and MORRIS, C. J. O. R. 1947. The precision of plasma determinations by the Evans Blue method. Brit. J. Exper. Path., 28:12–24.

References

MILLER, L. C. 1944. The U.S.P. collaborative digitalis study using frogs (1939–1941). J. Am. Pharm. A., 33:245–66.

MOORE, W., and BLISS, C. I. 1942. A method for determining insecticidal effectiveness using *Aphis rumicis* and certain organic compounds. J. Econ. Entomol., 35:544–53.

MORRELL, C. A., and ALLMARK, M. G. 1941. The toxicity and trypanocidal activity of commercial neoarsphenamine. J. Am. Pharm. A., 30:33–38.

NATIONAL BUREAU OF STANDARDS. 1950. Tables of the binomial probability distribution. Washington, D.C.: Government Printing Office.

PLACKETT, R. L., and BURMAN, J. P. 1946. The design of optimum multifactorial experiments. Biometrika, 33:305–25.

POTTER, C., and GILLHAM, E. M. 1946. Effects of atmospheric environment, before and after treatment, on the toxicity to insects of contact poisons. Ann. Appl. Biol., 33:142–59.

PRICE, W. C. 1946. Measurement of virus activity in plants. Biometrics, 2:81–86.

ROTHAMSTED EXPERIMENTAL STATION. 1938. Report for 1938.

SCHILD, H. O. 1942. A method of conducting a biological assay on a preparation giving repeated graded doses illustrated by the estimation of histamine. J. Physiol., 101:115–30.

SEWARD, E. H. 1949. Self-administered analgesia in labour. Lancet, 257: 781–83.

SOMERS, G. F. 1950. The measurement of thyroidal activity. Analyst 75:537–41.

SPENCER, E. L., and PRICE, W. C. 1943. Accuracy of the local-lesion method for measuring virus activity. I. Tobacco-mosaic virus. Am. J. Bot., 30:280–90.

STERN, C. 1950. Principles of human genetics. San Francisco: W. H. Freeman & Co.

TISCHER, R. G., and KEMPTHORNE, O. 1951. Influence of variations in technique and environment on the determination of consistency of canned sweet corn. Food Technol., 5:200–203.

WADLEY, F. M. 1948. Experimental design in comparison of allergens on cattle. Biometrics, 4:100–108.

WOOD, E. C. 1946. The theory of certain analytical procedures, with particular reference to micro-biological assays. Analyst, 71:1–14.

YATES, F. 1935. Complex experiments. J. Roy. Statist. Soc., Suppl., 2: 181–247.

———. 1952. Principles governing the amount of experimentation in developmental work. Nature, 170:138–40.

YOUDEN, W. J. 1937. Use of incomplete block replications in estimating tobacco-mosaic virus. Contr. Boyce Thompson Inst., 9:41–48.

Index

Adjustment for mean, 60
Agricultural research, 45, 82–83, 143
Alias, 96, 101
Alpine poppy, 9
Analgesia, 25–26, 69
Analysis of covariance, 39, 81, 157–60
Analysis of variance, 7, 53, 59, 79, 87, 91–92, 103, 156–57
Analytical bioassay, 123
Antibiotics, 129, 131
Anticoagulant therapy, 18, 21, 26
Aphis rumicis, 71

Bacillus subtilis, 126
Balanced incomplete blocks, 71–74, 77, 105, 107, 130, 136, 153
Balanced lattice, 77
Balanced lattice square, 78
Bias, 23, 32, 47–48, 128
Binomial distribution, 14, 19, 34, 40
Block, 50, 68–69, 129, 147
Blood pressure, 159
Blood sugar, 126, 133

Calculating machine, 34, 159–60
χ^2, 14–16, 20, 34
Chick method, 127
Classification, 2, 9, 29, 33, 42, 139
Clinical experiment, 22, 24–28, 118–20, 150
Completely randomized design, 49, 89
Concomitant measurement, 157–60
Confounding, 99–109, 131, 136, 146, 148, 153; double, 109; partial, 106–7, 132, 137, 146
Constraint, 63, 140
Contingency table, 7, 18

Continuity correction, 15, 20–21
Continuous variate, 29, 55, 138
Control, 18, 25, 31, 81, 149–50
Covariance analysis, 39, 81, 157–60
Cow, 52, 122
Cross-over design, 133, 137
Cubic lattice, 78
Cyclic permutation, 73
Cylinder-plate technique, 129, 131

Degrees of freedom (d.f.), 35, 41, 60, 91–92
Demonstration trial, 151
Dependent variate, 157
Digitalis, 141
Discrete variate, 9, 29, 55, 138
Double confounding, 109
Drosophila melanogaster, 85–88

Economy of research, 5, 30, 80, 83, 112, 117, 134, 140, 142–61
Efficiency, 17, 43–44, 85
Error, 53, 61, 94
Escherichia coli, 58
Estimation, 15, 35, 41–44, 47, 115–17, 120–22, 124, 128, 130–39, 156
Ethics, medical, 22, 27, 118, 150
Evans Blue, 54
Expectation, 10, 20
Experimental technique, 52, 54, 141, 157
Experimental unit, 3, 46, 107

Factorial design, 82–112, 114–15, 145, 153; 3^n, 88, 94, 99, 103–6, 110, 148; 2^n, 88, 91–92, 94, 97–99, 102–5, 110, 148
Fertility trend, 57
Fiducial limits, 42

167

Index

Fiducial probability, 42
5-point assay, 136
4-point assay, 129, 132
Fractional replication, 95–105, 109–12, 114–15, 146, 148, 153
Frequency distribution, 10

Gene frequency, 16
Generalized product rule, 97, 102
Genetics, 9, 30, 120, 140
Graeco-Latin square, 63–64, 77
Guinea pig, 57, 132

Headache, 25–26
Histaminase, 64
Histamine, 132

Independent variate, 157
Industrial research, 110, 115, 118
Insecticidal toxicity, 71, 89–90, 138
Insulin, 126, 133
Interaction, 46, 83, 91–110, 112, 114
Interblock information, 79–80
Intrablock information, 79

Lactobacillus helveticus, 124
Latin cube, 67
Latin square, 25, 49, 56–67, 75–78, 100, 109, 129, 132, 136, 153; orthogonal partition of, 64
Lattice, 76–78, 147, 153
Lattice square, 78, 147
Level, 88–90, 145–49
Limits of error, 15

Main effect, 91–92, 95–96
Main plot, 107–9
Median effective dose (ED50), 121, 138
Median lethal concentration, 72, 90
Mendelian ratios, 10, 17
Microbiological assay, 129–30
Mixed factorial design, 89, 99, 106, 148
Model, 11–13, 19, 142
Myocardial infarction, 18, 26

Neoarsphenamine, 141
Nicotine, 71
Normal distribution, 37–38, 40

Null hypothesis, 19, 33, 37, 54, 61
Number of experiments, 153–56
Nutrition experiments, 122

Observation plots, 151
Oestrone, 129
Optimal conditions, 115–17
Orthogonality, 62–63, 69, 76, 79, 92, 100

Pain, 69, 72
Paired observations, 39, 50, 118
Parallel line assay, 126, 129–36, 138, 140
Parameter, 124
Partial confounding, 106–7, 132, 137, 146
Partially balanced incomplete blocks, 78–79, 153
Penetrance, 16–17
Penicillin, 72, 115
Pilot experiment, 135
Placebo, 25
Plaid square, 109
Plasma volume, 54
Plot, 3, 46, 52, 57, 68, 114
Poisson distribution, 55
Positional effects, 50, 58, 130
Precision, 43–45, 49, 80, 85, 107, 110–12, 128, 133–41, 156–60
Probability, 11–17, 33–34, 37, 152; fiducial, 42
Pyrethrins, 89

Quantal response, 138, 141
Quasi-factor, 89, 101

Rabbit, 59, 126, 133
Random mating, 16
Random-number tables, 32, 48
Randomization, 23, 27, 32, 39–41, 47–49, 58, 65, 74–77, 102, 151, 153
Randomized blocks, 51–55, 81, 89, 100, 129, 132, 136, 153
Rat, 31, 129, 141
Rectangular lattice, 78
Red blood cells, 52, 55
Regression equation, 124
Relative potency, 90, 124–41

Index

Replication, 47, 76, 80, 92–95, 111, 145–46, 150–52, 158
Residual effects, 66
Response, 124, 128, 135, 138, 140–41
Riboflavin, 124

Sampling, 156
Segregation, genetic, 9, 30, 120
Sequential design, 113–22
Significance test, 13–14, 21, 37, 43, 54, 61–62, 87
Single-replicate design, 94, 103
6-point assay, 131
Slope ratio assay, 125, 136–38, 140
Southern bean mosaic, 131
Split-plot design, 107–9
Staircase estimation, 120–22
Standard deviation, 35
Standard error, 17, 36, 40, 43, 54, 151
Standard preparation, 123, 135, 137
Standard response curve, 126–28, 135
Streptomycin, 126
Subplot, 108–9
Sugar-beet, 84
Sugar cane, 105
Sum of squares, 34, 59–60

t-test, 37, 40–42, 54
Test preparation, 123, 135, 137
Testicular diffusing factor, 59
3-point assay, 136
Thyrotrophin, 57
Tobacco mosaic virus, 74
Treatment, 3, 46, 80, 89, 144–45
Tribolium castaneum, 89
Trypanocide, 141
Tuberculin, 52

Validity test, 131–37
Variance, 17, 35, 38, 53, 151
Variance analysis, 7, 53, 59, 79, 87, 91–92, 103, 156–57
Variance homogeneity, 38
Virus inoculation, 57, 71, 74, 101, 130
Vitamin A, 31
Vitamin B_{12}, 58
Vitamin D, 127
Vitamin E, 31

Weighing, 109–10

Yates's correction, 15, 20–21
Youden square, 74–76, 137, 153